Immersive Audio Signal Processing

Information Technology: Transmission, Processing, and Storage

Series Editors: **Robert Gallager**
Massachusetts Institute of Technology
Cambridge, Massachusetts

Jack Keil Wolf
University of California, San Diego
La Jolla, California

Immersive Audio Signal Processing

Sunil Bharitkar

Audyssey Laboratories, Inc. and

University of Southern California

Los Angeles, CA, USA

Chris Kyriakakis

University of Southern California

Los Angeles, CA, USA

 Springer

Sunil Bharitkar
Dept. of Electrical Eng.-Systems
University of Southern California
Los Angeles, CA 90089-2564, and
Audyssey Laboratories, Inc.
350 S. Figueroa Street, Ste. 196
Los Angeles, CA 90071
sunil@audyssey.com

Chris Kyriakakis
Dept. of Electrical Eng.-Systems and
Integrated Media Systems Center (IMSC)
University of Southern California
Los Angeles, CA 90089-2564
ckyriak@imsc.usc.edu

ISBN 978-1-4419-2105-5   e-ISBN 978-0-387-28503-0
e-ISBN: 0-387-28503-2

Printed on acid-free paper.

9 8 7 6 5 4 3 2 1

springer.com

To Aai and Baba

To Wee Ling, Anthony, and Alexandra

Preface

This book is the result of several years of research in acoustics, digital signal process-

This book is the result of several years of research in acoustics, digital signal processing (DSP), and psychoacoustics, conducted in the Immersive Audio Laboratory at the
University of Southern California's (USC) Viterbi School of Engineering, the Signal
and Image Processing Institute at USC, and Audyssey Laboratories. The authors'
association began five years ago when Sunil Bharitkar joined Chris Kyriakakis' research group as a PhD candidate.

The title *Immersive Audio Signal Processing* refers to the fact that signal processing algorithms that take into account human perception and room acoustics, as
described in this book, can greatly enhance the experience of immersion for listeners.
The topics covered are of widespread interest in both consumer and professional audio, and have not been previously presented comprehensively in an audio processing
textbook.

Besides the basics of DSP and psychoacoustics, this book contains the latest
results in audio processing for audio synthesis and rendering, multichannel room
equalization, audio selective signal cancellation, signal processing for audio applications, surround sound synthesis and processing, and the incorporation of psychoacoustics in audio signal processing algorithms.

Chapter 1, "Foundations of Digital Signal Processing for Audio," includes concepts from signals and linear systems, analog–to–digital and digital–to–analog conversion, convolution, digital filtering concepts, sampling rate alteration, and transfer
function representations (viz., z-transforms, Fourier transforms, bilinear transforms).

Chapter 2, "Filter Design Techniques for Audio Processing," introduces the design of various filters such as FIR, IIR, parametric, and shelving filters.

Chapter 3, "Introduction to Acoustics and Auditory Perception," introduces the
theory and physics behind sound propagation in enclosed environments, room acoustics, reverberation time, the decibel scale, loudspeaker and room responses, and
stimuli for measuring room responses (e.g., logarithmic chirp, maximum length sequences). We also briefly discuss some relevant topics in pyschoacoustics, such as
loudness perception and frequency selectivity.

In Chapter 4, "Immersive Audio Synthesis and Rendering," we present techniques that can be used to automatically generate multiple microphone signals

needed for a multichannel rendering without having to record using multiple real microphones for performing spatial audio playback over loudspeakers. We also present techniques for spatial audio playback over loudspeakers. It is assumed that readers have sufficient knowledge of head-related transfer functions (HRTFs). However, adequate references are provided at the end of the book for interested readers.

Chapter 5, "Multiple Position Room Response Equalization for Real-Time Applications," provides the necessary theory behind equalization of room acoustics for immersive audio playback. Theoretical analysis and examples for single listener and multiple listener equalization are provided. Traditional techniques of single position equalization using FIR and IIR filters are introduced. Subsequently, a multiple listener equalization technique employing a pattern recognition technique is presented. For real-time implementations, warping for designing lower filter orders is introduced. The motivation for the pattern recognition approach can be seen through a statistical analysis and visual interpretation of the clustering phenomena through the Sammon map algorithm. The Sammon map also permits a visual display of room response variations as well as a multiple listener equalization performance measure. The influence of reverberation on room equalization is also discussed. Results from a typical home theater setup are presented in the chapter.

Chapter 6, "Practical Considerations for Multichannel Equalization," discusses distortions due to phase effects, and presents algorithms that minimize the effect of phase distortions. Selecting proper choices of bass management filters, crossover frequencies, as well as all-pass coefficients and time-delay adjustments that affect crossover region response are presented.

Chapter 7, "Robustness of Equalization to Displacement Effects: Part I," explores robustness analysis (viz., mismatch between listener positions during playback and microphone position during room response measurement) in room equalization for frequencies above the Schroeder frequency.

Chapter 8, "Robustness of Equalization to Displacement Effects: Part II," explores robustness analysis in room equalization for low frequencies.

Chapter 9,"Selective Audio Signal Cancellation," presents a signal processing-based approach for audio signal cancellation at predetermined positions. This is important, for example, in automobile environments, to create a zone of silence spatially.

The material in this book is primarily intended for the practicing engineer, scientists, and researchers in the field. It is also suitable for a semester course at the upper-class undergraduate and graduate level. A basic knowledge of signal processing and linear system theory is assumed, although relevant topics are presented early on in this book. References to supplemental information are given at the end of the book.

Several individuals provided technical comments and insight on a preliminary version of the manuscript for improving the book. Hence, we would like to acknowledge and thank the following individuals, Dr. Randy Cole from Texas Instruments, Prof. Tomlinson Holman from the University of Southern California, and Prof. Stephan Weiss from the University of Southampton. Ana Bozicevic and Vaishali Damle at Springer, provided incentive to the authors to produce the manuscript and make the book a reality, and we are thankful for their valuable assistance during the process. We would also like to thank Elizabeth Loew for production of this volume. Thanks also go out to the people at Audyssey Laboratories, in particular Philip Hilmes and Michael Solomon, for their support during the preparation of this manuscript.

We invite you to join us on this exciting journey, where signal processing, acoustics, and auditory perception have merged to create a truly immersive experience.

Los Angeles, California   *Sunil Bharitkar*
July, 2005   *Chris Kyriakakis*

Contents

Part II Acoustics and Auditory Perception

Part III Immersive Audio Processing

Part I

Digital Signal Processing for Audio and Acoustics

1

Foundations of Digital Signal Processing for Audio and Acoustics

The content presented in this chapter includes relevant topics in digital signal processing such as the mathematical foundations of signal processing (viz., convolution, sampling theory, etc.), basics of linear and time-invariant (LTI) systems, minimum-phase and all-pass systems, sampling and reconstruction of signals, discrete time Fourier transform (DTFT), discrete Fourier transform (DFT), z-transform, bilinear transform, and linear-phase finite impulse response (FIR) filters.

1.1 Basics of Digital Signal Processing

Digital signal processing (DSP) involves either one or more of the following [1]: (i) modeling or representation of continuous time signals (viz., analog signals) by discrete time or digital signals, (ii) operation on digital signals, typically through linear or nonlinear filtering and/or time to frequency mapping, to transform them to desirable signals, and (iii) the generation of continuous time signals from digital signals. The block diagram for a general DSP system is shown in Fig. 1.1 where the blocks identify each of the three processes described above. In this book, there is an implicit assumption of such DSP systems satisfying linearity and time-invariance (i.e., LTI property as explained below) unless explicitly stated.

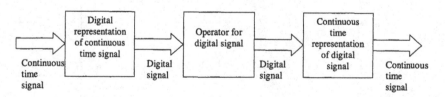

Fig. 1.1. General digital signal processing system.

1.1.1 Discrete Time Signals and Sequences

A discrete time signal, $x(n)$, is represented as a sequence of numbers x with corresponding time indices represented by integer values n [2]. In reality, such discrete time signals can arise out of sampling a continuous time signal (viz., the first block in Fig. 1.1). Specifically,

$$x(n) = x_c(nT_s) \tag{1.1}$$

where the continuous time signal $x_c(t)$ is sampled with a sampling period T_s which is the inverse of the sampling frequency f_s. Typical sampling frequencies used in audio processing applications include 32 kHz, 44.1 kHz, 48 kHz, 64 kHz, 96 kHz, and 192 kHz.

Some examples of discrete time signals include:

(i) Kronecker delta function shown in Fig. 1.2 and as defined by

$$x(n) = \delta(n) = \begin{cases} 1 & n = 0 \\ 0 & n \neq 0 \end{cases} \tag{1.2}$$

(ii) Exponential sequence shown in Fig. 1.3 and as defined by

$$x(n) = A\alpha^n \tag{1.3}$$

(iii) Step sequence shown in Fig. 1.4 and as defined by

$$x(n) = u(n) = \sum_{k=-\infty}^{n} \delta(k) = \begin{cases} 1 & n \geq 0 \\ 0 & n < 0 \end{cases} \tag{1.4}$$

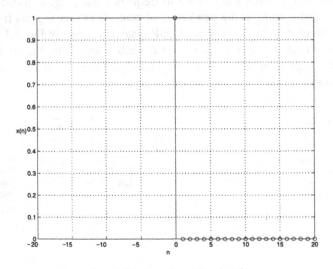

Fig. 1.2. Kronecker delta signal.

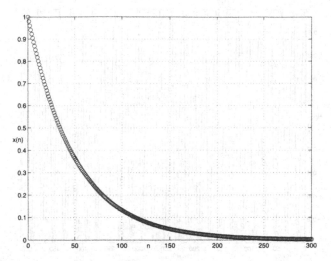

Fig. 1.3. Exponentially decaying sequence with $A = 1$ and $\alpha = 0.98$.

(iv) The special case when $\alpha = e^{j\omega_0 n}$ and $A = |A|e^{j\phi}$ the resulting magnitude and phase of the complex exponential sequence shown in Fig. 1.5 (in the equivalent continuous form) and given by

$$x(n) = Ae^{j(\omega_0 n + \phi)} = A(\cos(\omega_0 n + \phi) + j\sin(\omega_0 n + \phi)) \qquad (1.5)$$

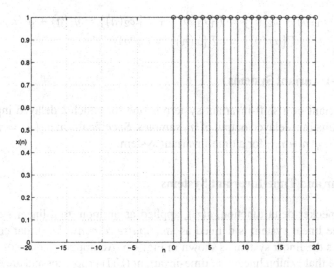

Fig. 1.4. Step sequence.

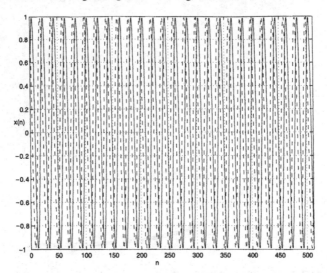

Fig. 1.5. Real and imaginary part of a complex exponential sequence with $\phi = 0.25\pi$ and $\omega_0 = 0.1\pi$.

1.1.2 Linear Systems

Operations performed by a DSP system are based on the premise that the system satisfies the properties of linearity and time-invariance. Specifically, from the theory of linear systems, if $T\{.\}$ denotes the transformation performed by a linear system (i.e., $y(n) = T\{x(n)\}$), then the input and output of a linear system satisfy the following properties of additivity and homogeneity, respectively,

$$T\{x_1(n) + x_2(n)\} = T\{x_1(n)\} + T\{x_2(n)\} = y_1(n) + y_2(n)$$
$$T\{ax(n)\} = aT\{x(n)\} = ay(n) \tag{1.6}$$

1.1.3 Time-Invariant Systems

A time-invariant or a shift-invariant system is one for which a delayed input of n_0 samples results in a delayed output of n_0 samples. Specifically, if $x_1(n) = x(n-n_0)$ then $y_1(n) = y(n - n_0)$ for a time-invariant system.

1.1.4 Linear and Time-Invariant Systems

With a Kronecker delta function, $\delta(n)$, applied as an input to a linear system, the output of the linear system is defined as an *impulse response* $h(n)$ that completely characterizes the linear system as shown in Fig. 1.6. An important class of DSP systems is those that exhibit linear *and* time-invariant (LTI) properties and are extremely important for designing immersive audio signal processing systems. Thus, if the input $x(n)$ is represented as a series of delayed impulses as

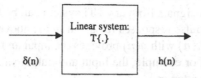

Fig. 1.6. A linear system with impulse response $h(n)$.

$$x(n) = \sum_{k=-\infty}^{\infty} x(k)\delta(n-k) \tag{1.7}$$

then the output $y(n)$ can be expressed as the well-known convolution formula where $h(n)$ is the impulse response,

$$
\begin{aligned}
y(n) &= T\left\{ \sum_{k=-\infty}^{\infty} x(k)\delta(n-k) \right\} \\
&= \sum_{k=-\infty}^{\infty} x(k)T\{\delta(n-k)\} \\
&= \sum_{k=-\infty}^{\infty} x(k)h(n-k) \\
&= \sum_{k=-\infty}^{\infty} h(k)x(n-k) \\
&= x(n) \otimes h(n) \tag{1.8}
\end{aligned}
$$

Some common examples of LTI systems include:
(i) Ideal Delay

$$h(n) = \delta(n-n_d) \tag{1.9}$$

(ii) Moving Average (MA)

$$h(n) = \frac{1}{M_1 + M_2 + 1} \sum_{k=M_1}^{M_2} \delta(n-k) \tag{1.10}$$

(iii) Autoregressive (AR)

$$\sum_{i=N_1}^{N_2} a_i h(n-i) = \delta(n) \tag{1.11}$$

(iv) Autoregressive and Moving Average (ARMA)

$$\sum_{i=N_1}^{N_2} a_i h(n-i) = \sum_{k=M_1}^{M_2} b_i \delta(n-k) \tag{1.12}$$

The input and output signals from any LTI system can be found through various methods. For simple impulse responses (as given in the above examples), substituting $\delta(n)$ with $x(n)$ and $h(n)$ with $y(n)$ provides the input and output signal description of the LTI system. For example, the input and output signal description for the ARMA system can be written as

$$\sum_{i=N_1}^{N_2} a_i y(n-i) = \sum_{k=M_1}^{M_2} b_i x(n-k) \tag{1.13}$$

A second method for finding the output from an LTI system involves the convolution formula (1.8) (along with any graphical plotting). For example, if $x(n) = \alpha_1^n u(n)$ and $h(n) = \alpha_2^n u(n)$ are two right-sided sequences (where $|\alpha_1| < 1$ and $|\alpha_2| < 1$ and $u(n)$ is the step function), then with (1.8),

$$y(n) = x(n) \otimes h(n)$$

$$= \sum_{k=0}^{n} \alpha_1^k \alpha_2^{n-k}$$

$$= \alpha_2^n \sum_{k=0}^{n} (\frac{\alpha_1}{\alpha_2})^k$$

$$= \frac{\alpha_2^{n+1} - \alpha_1^{n+1}}{\alpha_2 - \alpha_1} \tag{1.14}$$

where the following series expansion has been used

$$\sum_{k=N_1}^{N_2} \alpha^k = \frac{\alpha^{N_1} - \alpha^{N_2+1}}{1-\alpha} \tag{1.15}$$

MATLAB software, from Mathworks, Inc. (http://www.mathworks.com), includes the conv(a,b) command for convolving two signals. An example from using this function with $\alpha_1 = 0.3$ and $\alpha_2 = 0.6$ is shown in Fig. 1.7. Finally, another method for determining the output signal is through the use of linear transform theory (viz., the Fourier transform and the z-transform).

1.2 Fourier Transforms

Complex exponential sequences are eigenfunctions of LTI systems and the response to a sinusoidal input is a sinusoid with the same frequency of the input signal and with the amplitude and phase as determined by the system [2]. Specifically, when $x(n) = e^{j\omega n}$, then the output $y(n)$ can be expressed by the convolution formula (1.8) as

$$y(n) = \sum_{k=-\infty}^{\infty} h(k) e^{j\omega(n-k)}$$

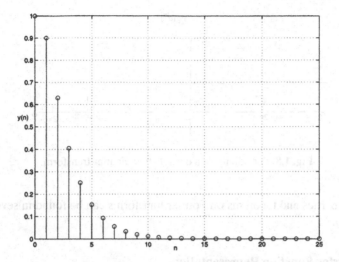

Fig. 1.7. The convolution of two exponentially decaying right-sided sequences (viz., (1.14)) with $\alpha_1 = 0.3$ and $\alpha_2 = 0.6$.

$$= e^{j\omega n} \sum_{k=-\infty}^{\infty} h(k)e^{-j\omega k}$$

$$= e^{j\omega n} H(e^{j\omega}) \qquad (1.16)$$

where $H(e^{j\omega}) = H_R(e^{j\omega}) + jH_I(e^{j\omega}) = |H(e^{j\omega})|e^{j\angle H(e^{j\omega})}$ is the *discrete time Fourier transform* of the system and characterizes the LTI system along a real frequency axis ω expressed in radians. The real frequency variable ω is related to the analog frequency $\Omega = 2\pi f$ (f is in Hz), as shown in a subsequent section on sampling, through the relation $\omega = \Omega T_s$ where the sampling frequency f_s and the sampling period T_s are related by $T_s = 1/f_s$.

An important property of the discrete time Fourier transform or frequency response, $H(e^{j\omega})$, is that it is periodic with a period of 2π, as

$$H(e^{j\omega}) = \sum_{k=-\infty}^{\infty} h(k)e^{-j\omega k} = \sum_{k=-\infty}^{\infty} h(k)e^{-j(\omega+2\pi)k} = H(e^{j(\omega+2\pi)}).$$

A periodic discrete time frequency response of a low-pass filter, with a cutoff frequency ω_c, is shown in Fig. 1.8.

Thus, the discrete time forward and inverse Fourier transforms can be expressed as

$$h(n) \longleftrightarrow H(e^{j\omega}) \qquad (1.17)$$

$$H(e^{j\omega}) = \sum_{k=-\infty}^{\infty} h(k)e^{-j\omega k} \qquad (1.18)$$

$$h(n) = \frac{1}{2\pi} \int_{2\pi} H(e^{j\omega})e^{j\omega n} d\omega \qquad (1.19)$$

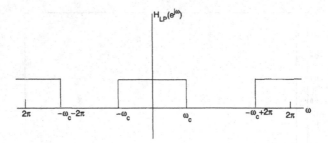

Fig. 1.8. Periodicity of a discrete time Fourier transform.

The properties and theorems on Fourier transforms can be found in several texts (e.g., [2, 3]).

1.2.1 Transfer Function Representation

In general, if an LTI system can be expressed as a ratio of numerator and denominator polynomials,

$$H(e^{j\omega}) = \frac{\sum_{k=1}^{M} a_k e^{-j\omega k}}{\sum_{k=1}^{N} b_k e^{-j\omega k}} = \frac{b_0 \prod_{k=1}^{M}(1 - c_k e^{-j\omega})}{a_0 \prod_{k=1}^{N}(1 - d_k e^{-j\omega})} \qquad (1.20)$$

and is stable (i.e., $|d_k| < 1 \forall k$) then the magnitude response of the transfer function in decibel scale (dB scale) can be expressed as

$$|H(e^{j\omega})| = \sqrt{H(e^{j\omega})H^*(e^{j\omega})}$$

$$= \frac{b_0}{a_0} \sqrt{\frac{\prod_{k=1}^{M}(1 - c_k e^{-j\omega}) \prod_{k=1}^{M}(1 - c_k^* e^{j\omega})}{\prod_{k=1}^{N}(1 - d_k e^{-j\omega}) \prod_{k=1}^{N}(1 - d_k^* e^{j\omega})}} \qquad (1.21)$$

$$|H(e^{j\omega})|(\mathrm{dB}) = 20 \left(\log_{10} \frac{b_0}{a_0} + \sum_{k=1}^{M} \log_{10}|1 - c_k e^{-j\omega}| - \sum_{k=1}^{N} \log_{10}|1 - d_k e^{-j\omega}| \right)$$

the phase response can be written as

$$\angle H(e^{j\omega}) = \angle \left(\frac{b_0}{a_0} \right) + \sum_{k=1}^{M} \angle(1 - c_k e^{-j\omega}) - \sum_{k=1}^{N} \angle(1 - d_k e^{-j\omega}) \qquad (1.22)$$

and the group delay is

$$\mathrm{grd}[H(e^{j\omega})] = -\frac{\partial}{\partial \omega} \angle H(e^{j\omega}) \qquad (1.23)$$

$$= -\sum_{k=1}^{M} \frac{\partial}{\partial \omega} \arg(1 - c_k e^{-j\omega}) + \sum_{k=1}^{N} \frac{\partial}{\partial \omega} \arg(1 - d_k e^{-j\omega})$$

The numerator roots c_k and the denominator roots d_k of the transfer function $H(e^{j\omega})$ (1.20) are called the zeros and poles of the transfer function, respectively.

Because $H(e^{j\omega}) = H(e^{j(\omega+2\pi)})$, the phase of each of the terms in the phase response (1.22) is ambiguous. A correct phase response can be obtained by taking the principal value $ARG(H(e^{j\omega}))$ (which lies between $-\pi$ and π), computed by any computer subroutine (e.g., the *angle* command in MATLAB) or the arctangent function from a calculator, and adding $2\pi r(\omega)$ [4]. Thus,

$$\angle H(e^{j\omega}) = ARG(H(e^{j\omega})) + 2\pi r(\omega) \tag{1.24}$$

The *unwrap* command in MATLAB computes the angle in terms of the principal value. For example, a fourth-order ($N = 4$) low-pass Butterworth transfer function used for audio bass management, with a cutoff frequency ω_c, can be expressed as

$$H(e^{j\omega}) = \prod_{k=0}^{\frac{N}{2}-1} \frac{b_{0,k} + b_{1,k}e^{-j\omega} + b_{2,k}e^{-j2\omega}}{a_{0,k} + a_{1,k}e^{-j\omega} + a_{2,k}e^{-j2\omega}} \tag{1.25}$$

$$b_{0,k} = b_{2,k} = K^2$$
$$b_{1,k} = 2K^2$$
$$a_{0,k} = 1 + 2K\cos(\pi(2k+1)/2N) + K^2$$
$$a_{2,k} = 1 - 2K\cos(\pi(2k+1)/2N) + K^2$$
$$a_{1,k} = 2(K^2 - 1)$$

where $K = \tan(\omega_c/2) = \tan(\pi f_c/f_s)$. The magnitude response, principal phase, and unwrapped phase, for the fourth-order Butterworth low-pass filter with cutoff frequency $f_c = 80$ Hz and $f_s = 48$ kHz, as shown in Fig. 1.9, reveal the 2π phase rotation in principal value (viz., Fig. 1.9(b)).

Minimum-Phase Systems

From system theory, if $H(e^{j\omega})$ is assumed to correspond to a causal[1] and stable system, then the magnitude of all its poles is less than unity [2]. For certain classes of problems it is important to constrain the inverse of the transfer function, $H(e^{j\omega})$, to be causal and stable. Hence if $H(e^{j\omega})$ is causal and stable, then if the inverse, $1/H(e^{j\omega})$, too is constrained to be causal and stable it must be true that the magnitude of all of the zeros (i.e., the roots of the numerator polynomial) of $H(e^{j\omega})$ must be less than unity. Classes of systems that satisfy this property (where the transfer function, as well as its inverse, is causal and stable) are called *minimum-phase* systems [2] and the transfer function is usually denoted by $H_{\min}(e^{j\omega})$.

---

[1] A causal system is one for which the output signal depends on the present value and/or the past values of the input signal or $y(n) = f[x(n), x(n-1), \ldots, x(n-p)]$, where $p \geq 0$.

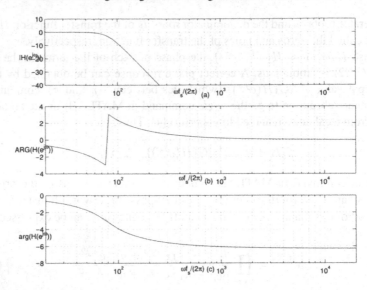

Fig. 1.9. (a) Magnitude response of the fourth-order Butterworth low-pass filter; (b) principal value of the phase; (c) unwrapped phase.

All-Pass Systems

An all-pass system or transfer function is one whose magnitude response is flat for all frequencies. A first-order all-pass transfer function can be expressed as

$$H_{ap}(e^{j\omega}) = \frac{e^{-j\omega} - \lambda^*}{1 - \lambda e^{-j\omega}} = e^{-j\omega}\frac{1 - \lambda^* e^{j\omega}}{1 - \lambda e^{-j\omega}} \tag{1.26}$$

where the roots of the numerator and denominator (viz., $1/\lambda^*$ and λ, respectively) are conjugate reciprocals of each other. Thus,

$$|H_{ap}(e^{j\omega})| = \sqrt{H_{ap}(e^{j\omega})H_{ap}^*(e^{j\omega})}$$

$$= \sqrt{e^{-j\omega}\frac{1 - \lambda^* e^{j\omega}}{1 - \lambda e^{-j\omega}}e^{j\omega}\frac{1 - \lambda e^{-j\omega}}{1 - \lambda^* e^{j\omega}}} = 1 \tag{1.27}$$

A generalized all-pass transfer function, providing a real time-domain response, can be expressed as [2],

$$H_{ap}(e^{j\omega}) = \prod_{i=1}^{N_{real}} \frac{e^{-j\omega} - d_i}{1 - d_i e^{-j\omega}} \prod_{k=1}^{N_{complex}} \frac{e^{-j\omega} - g_k^*}{1 - g_k e^{-j\omega}}\frac{e^{-j\omega} - g_k}{1 - g_k^* e^{-j\omega}} \tag{1.28}$$

where d_i is a real pole and g_k is a complex pole.

The phase responses for a first-order and second-order all-pass transfer function are [2],

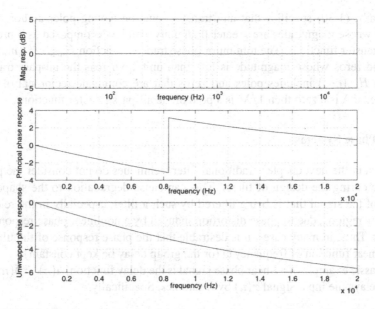

Fig. 1.10. (a) Magnitude response of a second order real response all-pass filter; (b) principal value of the phase; (c) unwrapped phase.

$$\angle \frac{e^{-j\omega} - re^{-j\theta}}{1 - re^{j\theta}e^{-j\omega}} = -\omega - 2\arctan\frac{r\sin(\omega - \theta)}{1 - r\cos(\omega - \theta)}$$

$$\angle\left[\left(\frac{e^{-j\omega} - re^{-j\theta}}{1 - re^{j\theta}e^{-j\omega}}\right)\left(\frac{e^{-j\omega} - re^{j\theta}}{1 - re^{-j\theta}e^{-j\omega}}\right)\right] = -2\omega - 2\arctan\frac{r\sin(\omega - \theta)}{1 - r\cos(\omega - \theta)}$$

$$-2\arctan\frac{r\sin(\omega + \theta)}{1 - r\cos(\omega + \theta)} \quad (1.29)$$

The magnitude and phase response for a second-order all-pass transfer function with complex poles ($r = 0.2865$ and $\theta = 0.1625\pi$) is shown in Fig. 1.10.

As shown in a subsequent chapter, using an all-pass filter in cascade with other filters allows the overall phase response of a system to approximate a desired phase response which is useful to correct for phase interactions between the subwoofer and satellite speaker responses in a multichannel sound playback system.

Decomposition of Transfer Functions

An important fact on any rational transfer function is that it can be decomposed as a product of two individual transfer functions (viz., the minimum-phase function and the all-pass function) [2]. Thus,

$$H(e^{j\omega}) = H_{\min}(e^{j\omega})H_{ap}(e^{j\omega}) \quad (1.30)$$

$$h(n) = h_{\min}(n) \otimes h_{ap}(n)$$

Specifically, (1.30) specifies that any transfer function having poles and/or zeros, some of whose magnitudes are greater than unity, can be decomposed as a product of two transfer functions. The minimum-phase transfer function $H_{\min}(e^{j\omega})$ includes poles and zeros whose magnitude is less than unity, whereas the all-pass transfer function $H_{ap}(e^{j\omega})$ includes poles and zeros that are conjugate reciprocal of each other (i.e., if λ is a zero then $1/\lambda^*$ is a pole of the all-pass transfer function).

Linear Phase Systems

As shown in the next chapter, traditional filter techniques do not consider the phase response during the design of filters. This can cause degradation to the shape and quality of the signal that is being filtered by such a filter, especially in the relevant frequency regions, due to phase distortion induced by a nonlinear phase response of the filter. Thus, in many cases, it is desirable that the phase response of the filter be kept a linear function of frequency ω (or the group delay be kept constant).

A classic example of a linear phase signal is the delay function, $h(n) = \delta(n-k)$, which delays the input signal $x(n)$ by k samples. Specifically,

$$
\begin{aligned}
y(n) &= h(n) \otimes x(n) = x(n-k) \\
H(e^{j\omega}) &= e^{-j\omega k} \\
|H(e^{j\omega})| &= 1 \\
\angle H(e^{j\omega}) &= -k\omega \\
\mathrm{grd}[H(e^{j\omega})] &= k
\end{aligned}
\tag{1.31}
$$

A *generalized linear phase* system function, $H(e^{j\omega})$, may be expressed as a product of a real function $A(e^{j\omega})$, with a complex exponential $e^{-j\psi}$ such that

$$
\begin{aligned}
H(e^{j\omega}) &= A(e^{j\omega})e^{-j\alpha\omega+j\beta} \\
\angle H(e^{j\omega}) &= \beta - \alpha\omega \\
\mathrm{grd}[H(e^{j\omega})] &= \alpha
\end{aligned}
\tag{1.32}
$$

More details on designing linear phase filters are given in the next chapter.

1.3 The z-Transform

A generalization of the Fourier transform is the z-Transform which can be expressed as a two-sided power series (also called the *Laurent series*), of a signal $x(n)$, in the complex variable $z = e^{j\omega}$ is expressed as

$$
X(z) = \sum_{n=-\infty}^{\infty} x(n)z^{-n}
\tag{1.33}
$$

whereas the inverse z-transform can be expressed in terms of the *contour integral*

$$x(n) = \frac{1}{2j\pi} \oint_C X(z)z^{n-1}dz \qquad (1.34)$$

A principle motivation for using this transform is that the Fourier transform does not converge for all discrete time signals, or sequences, and a generalization via the z-transform encompasses a broader class of signals. Furthermore, powerful complex variable techniques can be used to analyze signals when using the z-transform.

Being a complex variable, the poles and zeros of the resulting system function can be depicted on a two-dimensional complex z-plane of Fig. 1.11. Because the z-transform is related to the Fourier transform through the transformation $z = e^{j\omega}$, then the Fourier transform exists on the *unit circle* depicted in Fig. 1.11.

The region of convergence (ROC) is defined to be the set of values on the complex z-plane where the z-transform converges.

Some examples of using z-transforms for determining system functions corresponding to discrete time signals are given below.

(i) $x(n) = \delta(n - n_d) \implies X(z) = \sum_{n=-\infty}^{\infty} \delta(n - n_d)z^{-n} = z^{-n_d}$, where the ROC is z-plane.

(ii) $x(n) = a^n u(n) \implies X(z) = \sum_{n=0}^{\infty} a^n z^{-n} = \sum_{n=0}^{\infty}(az^{-1})^n = 1/(1 - az^{-1})$, where the ROC is the region $|z| > |a|$ exterior to the dotted circle in Fig. 1.12 (where $|a| = 0.65$ and the ROC includes the unit circle).

(iii) $x(n) = \begin{cases} (1/3)^n & n \geq 0 \\ 2^n & n < 0 \end{cases}$

$$X(z) = \sum_{n=-\infty}^{-1} (2z^{-1})^n + \sum_{n=0}^{\infty}((1/3)z^{-1})^n$$

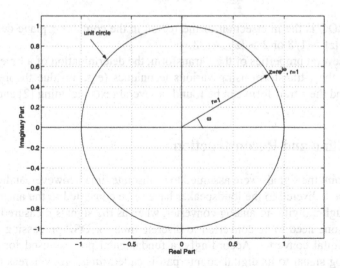

Fig. 1.11. The complex z-plane.

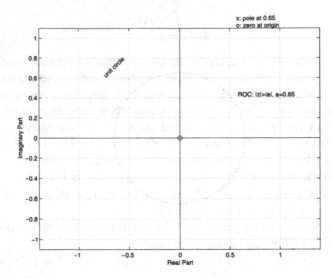

Fig. 1.12. The ROC in the complex z-plane for the sequence $x(n) = a^n u(n)$ with $a = 0.65$.

$$= \sum_{n=1}^{\infty} (2z^{-1})^{-n} + \sum_{n=0}^{\infty} ((1/3)z^{-1})^n$$

$$= \left(\sum_{n=0}^{\infty} ((1/2)z)^n - 1 \right) + \sum_{n=0}^{\infty} ((1/3)z^{-1})^n$$

$$= \frac{\frac{1}{2}z^{-1}}{1 - \frac{1}{2}z^{-1}} + \frac{1}{1 - \frac{1}{3}z^{-1}}$$

where the ROC is the intersection of the region in the complex z-plane defined by $|z| < 2$ and $|z| > 1/3$ or $2 > |z| > 1/3$.

Again, several properties of the z-transform, the determination of a time domain signal from the z-transform using various techniques (e.g., residue theorem), and theory behind the z-transform can be found in several texts including [2] and [4].

1.4 Sampling and Reconstruction

Up to this point the signals were assumed to be discrete time. However, audio signals that are to be delivered to the loudspeaker have to be converted to the analog counterpart through a digital-to-analog converter, whereas the signals measured through the microphone need to be converted to discrete time for DSP processing using an analog-to-digital converter. Accordingly, a fundamental process used for converting an analog signal to its digital counterpart is called *sampling*, whereas the basic process for converting a discrete time signal to its analog counterpart is called *reconstruction*.

1.4.1 Ideal Sampling

A continuous time to discrete time conversion is achieved through ideal sampling where a periodic pulse train (shown in Fig. 1.13),

$$p(t) = \sum_{k=-\infty}^{\infty} \delta(t - kT_s) \tag{1.35}$$

of sampling period $T_s = 1/f_s$ is multiplied with a continuous time signal $x(t)$ to obtain the sampled version of $x_s(t)$ given as

$$x_s(t) = x(t)p(t) = x(t) \sum_{k=-\infty}^{\infty} \delta(t - kT_s) = \sum_{k=-\infty}^{\infty} x(kT_s)\delta(t - kT_s) \tag{1.36}$$

Based on the properties of Fourier transform (viz., multiplication in the time domain is equivalent to convolution in the frequency domain), the continuous time frequency response, $X_s(j\Omega)$, of the sampled signal $x_s(t)$ can be expressed as

$$X_s(j\Omega) = \frac{1}{2\pi}X(j\Omega) \otimes P(j\Omega) \tag{1.37}$$

$$= \frac{1}{T_s} \sum_{k=-\infty}^{\infty} X(j\Omega - jk\Omega_s)$$

where $\Omega = 2\pi f$ is the continuous time angular frequency in rad/s, $\Omega_s = 2\pi f_s$ ($f_s = 1/T_s$), and $P(j\Omega) = (2\pi/T_s)\sum_{k=-\infty}^{\infty} \delta(j\Omega - jk\Omega_s)$.

Thus, 1.37 represents a periodicity in the frequency domain, upon sampling, such as shown in Fig. 1.14 for a bandlimited signal $x(t)$ with a limiting or cutoff frequency of Ω_c. From the figure, it can be observed that the signal $x(t)$ can be obtained by

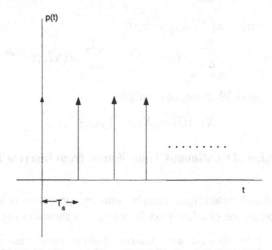

Fig. 1.13. The periodic pulse train of period $T_s = 125$ μs.

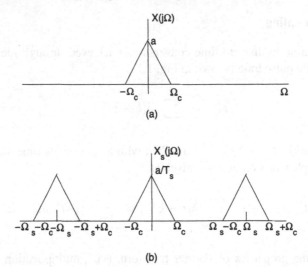

Fig. 1.14. (a) Fourier transform of a bandlimited signal $x(n)$ with limiting frequency Ω_c; (b) periodicity of the Fourier transform of the signal $x(n)$ upon ideal sampling.

simply low-pass filtering the baseband spectrum of $X_s(j\Omega)$ with a cutoff frequency Ω_c and inverse Fourier transforming the result. For recovering the signal $x(t)$, as can be seen from Fig. 1.14(b), it is required that $\Omega_s - \Omega_c > \Omega_c$ or $\Omega_s > 2\Omega_c$ to prevent an *aliased* signal recovery. This condition is called the *Nyquist condition*, Ω_c is called the *Nyquist frequency*, and Ω_s is called the *Nyquist rate*.

Subsequently, the frequency response of the discrete time signal from the sampled signal can be obtained from (1.36) by using the continuous time Fourier transform relation,[2]

$$X_s(j\Omega) = \sum_{k=-\infty}^{\infty} x(kT_s)e^{-jkT_s\Omega} \tag{1.38}$$

$$x(n) = x(t)|_{t=nT_s} = x(nT_s)$$

$$X(e^{j\omega}) = \sum_{n=-\infty}^{\infty} x(n)e^{-j\omega n} = \sum_{k=-\infty}^{\infty} x(nT_s)e^{-j\omega k} \tag{1.39}$$

Comparing (1.38) and (1.39) it can be seen that

$$X_s(j\Omega) = X(e^{j\omega})|_{\omega=\Omega T_s} \tag{1.40}$$

1.4.2 Reconstruction of Continuous Time Signals from Discrete Time Sequences

Reconstruction of bandlimited signals can be done by appropriately filtering the discrete time signal by means of a low-pass filter (e.g., as shown in Fig. 1.14(b)). This

---

[2] The continuous time forward and inverse Fourier transforms are $X(j\Omega) = \int_{-\infty}^{\infty} x(t)e^{-j\Omega t}dt$ and $x(t) = (1/2\pi)\int_{\infty}^{\infty} X(j\Omega)e^{j\Omega t}d\Omega$, respectively.

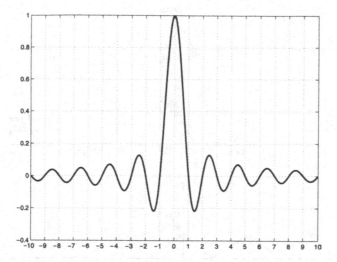

Fig. 1.15. The sinc interpolation filter for $T_s = 1$.

is mathematically described by

$$x_r(t) = \sum_{n=-\infty}^{\infty} x(nT_s)h_r(t - nT_s) = \sum_{n=-\infty}^{\infty} x(n)h_r(t - nT_s) \qquad (1.41)$$

By selecting $h_r(t)$ to be an ideal low-pass filter with a response of $h_r(t) = \sin(\pi t/T_s)/(\pi t/T_s)$,[3] (1.41) can be written as

$$x_r(t) = \sum_{n=-\infty}^{\infty} x(n)\frac{\sin(\pi(t - nT_s)/T_s)}{\pi(t - nT_s)/T_s} \qquad (1.42)$$

$$x_r(mT_s) = \sum_{n=-\infty}^{\infty} x(n)\frac{\sin(\pi(m - n))}{\pi(m - n)} = x(m) = x(mT_s)$$

because the sinc function is unity at time index zero and is zero at other discrete time indices as shown in Fig. 1.15 for $T_s = 1$. At other noninteger time values, the sinc filter acts as an interpolator by performing interpolation between the impulses of $x_s(t)$ to form the continuous time signal $x_r(t)$.

1.4.3 Sampling Rate Reduction by an Integer Factor

As shown in the previous section, a discrete time sequence can be obtained by sampling a continuous time signal $x(t)$ with a sampling frequency $f_s = 1/T_s$, and the subsequent sequence can be expressed as $x(n) = x(t)|_{t=nT_s} = x(nT_s)$. In many situations, it is necessary to reduce the sampling rate or frequency by an integer

---

[3] The function $\sin(\pi x)/(\pi x)$ is referred to as the sinc function.

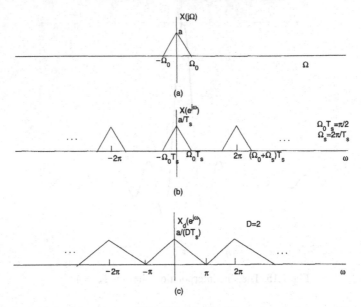

Fig. 1.16. (a) $x(t)$ having response $X(j\Omega)$ begin bandlimited such that $-\pi/D < \Omega T_s < \pi/D$; (b) $X(e^{j\omega})$; (c) $X_d(e^{j\omega})$ with $D = 2$.

amount.[4] Thus, in order to reduce the sampling rate by an amount D, the discrete time sequence is obtained by using a period T_s' such that $T_s' = DT_s \Rightarrow f_s' = f_s/D$, or $x_d(n) = x(nT_s') = x(nDT_s)$. The signal $x_d(n)$ is called a downsampled or decimated signal, which is obtained from $x(n)$ by reducing the sampling rate by a factor of D.

In order for $x_d(n)$ to be free of aliasing error, the continuous time signal $x(t)$ shown in Fig. 1.16(a), from which $x(n)$ is obtained, must be bandlimited a priori such that $-\pi/D < \Omega T_s < \pi/D$ or the original sampling rate should be at least D times the Nyquist rate.

The Fourier expression for the decimated signal $x_d(n)$ is

$$X_d(e^{j\omega}) = \sum_{k=-\infty}^{\infty} \frac{1}{DT_s} X\left(j\frac{\omega}{DT_s} - j\frac{2\pi k}{DT_s}\right) \qquad (1.43)$$

In time domain, $x_d(n)$ is obtained by removing every $D - 1$ samples from $x(n)$.

The block diagram for performing decimation is shown in Fig. 1.17, where the low-pass filter $H(e^{j\omega})$ (also called the anti-aliasing filter) is used for bandlimiting the signal $x(n)$ such that $\pi/D < \omega < \pi/D$ and the arrow indicates decimation.

---

[4] In audio applications, there are several sampling frequencies in use, including 32 kHz, 44.1 kHz, 64 kHz, 48 kHz, 96 kHz, 128 kHz, and 192 kHz, and in many instances it is required that the sampling rate be reduced by an integer amount, such as from 96 kHz to 48 kHz.

1.4.4 Increasing the Sampling Rate by an Integer Factor

In this case, the sampling rate increase is reflected by altering the sampling period such that $T_s' = T_s/L \Rightarrow f_s' = Lf_s$, or $x_i(n) = x(nT_s') = x(nT_s/L) = x(n/L)$, $n = 0, \pm L, \pm 2L, \ldots$. The signal $x_i(n)$ is called an interpolated signal, and is obtained from $x(n)$ by increasing the sampling rate by a factor of L. To obtain the interpolated signal, $x(n)$, the first step involves an expander stage [5, 6] that generates a signal $x_e(n)$ such that

$$x_e(n) = \begin{cases} x(n/L) & n = 0, \pm L, \pm 2L, \ldots \\ 0 & \text{otherwise} \end{cases}$$

$$x_e(n) = \sum_{k=-\infty}^{\infty} x(k)\delta(n - kL) \qquad (1.44)$$

The Fourier transform for the expander can be expressed as

$$X_e(e^{j\omega}) = \sum_{n=-\infty}^{\infty} x_e(n)e^{-j\omega n}$$

$$= \sum_{n=-\infty}^{\infty} \sum_{k=-\infty}^{\infty} x(k)\delta(n - kL)e^{-j\omega n}$$

$$= \sum_{k=-\infty}^{\infty} x(k)e^{-j\omega kL}$$

$$= X(e^{j\omega L}) \qquad (1.45)$$

Instead of generating zero valued samples every $L - 1$ samples, a better approach is to use to an interpolation stage using an ideal low-pass filter $h_i(n) = \sin(\pi n/L)/(\pi n/L)$, bandlimited between π/L and π/L, subsequent to the expander so that interpolated values are obtained for the intervening $L - 1$ samples. Thus, (1.44) becomes

$$x_i(n) = \sum_{k=-\infty}^{\infty} x(k)\frac{\sin(\pi(n - kL)/L)}{\pi(n - kL)/L} \qquad (1.46)$$

Fig. 1.18 shows the spectrum of a bandlimited continuous time signal $x(t)$ along with the expanded signal spectrum and the interpolated spectrum. As is evident from Fig. 1.18(c), the expander introduces $L - 1$ copies of the continuous time spectrum between $-\pi$ and π. Subsequently, an ideal low-pass interpolation filter, having a

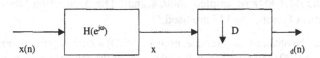

Fig. 1.17. System for performing decimation.

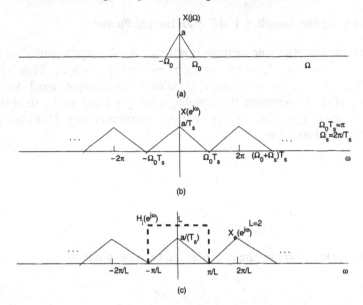

Fig. 1.18. (a) $x(t)$ having response $X(j\Omega)$; (b) $X(e^{j\omega})$; (c) extraction of baseband expanded and interpolated spectrum of $X(e^{j\omega})$ with $L = 2$.

cutoff frequency of π/L and a gain of L (shown by dotted lines in Fig. 1.18(c)), extracts the baseband discrete time interpolated spectrum of $X_e(j\Omega)$.

A block diagram employing an L-fold expander (depicted by an upwards arrow) and the interpolation filter $H_i(e^{j\omega})$ is shown in Fig. 1.19.

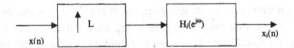

Fig. 1.19. System for performing interpolation.

1.4.5 Resampling for Audio Applications

Combining the decimation and interpolation processes, it is possible to obtain the various sampling rates used for audio processing. For example, if the original sampling rate that was used to convert the continuous time signal $x(t)$ to a discrete time sequence $x(n)$ was 48 kHz, and if it is required to generate an audio signal at 44.1 kHz (i.e., compact disc or CD-rate), then the block diagram of Fig. 1.20 may be used to generate the 44.1 kHz resampled audio signal. The decimation factor $D = 160$ and interpolation factor $L = 147$ are used.[5]

---

[5] This ratio can be obtained, for example, using the [N,D] = rat(X,tol) function in MATLAB, where $X = 44100/48000$ and tol is the tolerance for the approximation to determine the numerator and denominator integers.

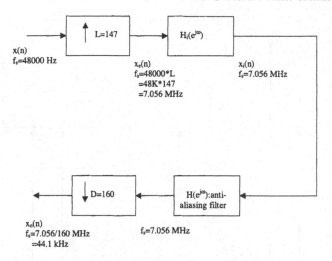

Fig. 1.20. System for performing resampling.

The function resample(x,A,B) in MATLAB (where A is the new sampling rate and B is the original sampling rate) also converts an audio signal $x(n)$ between different rates.

1.5 Discrete Fourier Transform

The forward (analysis) and inverse (synthesis) discrete Fourier transform (DFT) of a finite duration signal $x(n)$, of length N, are expressed as

$$X(k) = \sum_{n=-\infty}^{\infty} x(n)e^{(-jk2\pi n/N)} \quad : \quad \text{Analysis} \tag{1.47}$$

$$x(n) = \frac{1}{N} \sum_{n=-\infty}^{\infty} X(k)e^{(jk2\pi n/N)} \quad : \quad \text{Synthesis} \tag{1.48}$$

The relation between the DFT and the discrete time Fourier transform (DTFT) is

$$X(e^{j\omega})|_{\omega=2\pi k/N} = X(k), \quad k = 0, \dots, N-1 \tag{1.49}$$

Equation (1.49) basically states that the DFT is obtained by uniformly, or equally, sampling the DTFT (i.e., uniform sampling along the unit circle in the complex z-plane).

An important property for the DFT is the circular shift of an aperiodic signal, where any delay to a signal constitutes a circular shift in the signal. The appropriate relation between the DFT and the m-sample circularly shifted sequence is

$$x((n-m)_N) = x((n-m)\text{modulo}N) \leftrightarrow e^{-jk(2\pi/N)m}X(k) \tag{1.50}$$

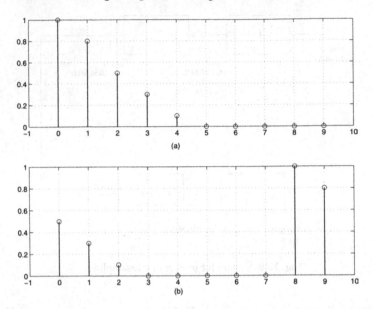

Fig. 1.21. Example of modulo shifting with $m = 2$.

An example of modulo shifting operation for $m = -2$ for the sequence $x(n)$ is shown in Fig. 1.21.

The N-point circular convolution of two finite length sequences, $x_1(n)$ and $x_2(n)$, each of length N, is expressed as

$$x_3(n) = x_1(n) \underset{N}{\odot} x_2(n) = \sum_{k=0}^{N-1} x_1(m)x_2((n-m)_N) \quad n = 0, \ldots, N-1$$

$$x_3(n) \leftrightarrow X_3(k) = X_1(k)X_2(k) \tag{1.51}$$

By simply considering each of the N length sequences, $x_1(n)$ and $x_2(n)$, as $2N$ length sequences (i.e., by appending N zeros to each sequence) the $2N$-point circular convolution of the augmented sequences is identical to the linear convolution of (1.8).

Again, properties of the DFT can be found in several texts including [2].

1.6 Bilinear Transform

Bilinear transform is used to convert between the continuous time and discrete time frequency variables. If $H_c(s)$ and $H(z)$ are the continuous time and discrete time frequency responses (where $s = \sigma + j\Omega$ and $z = e^{j\omega}$), the bilinear transform between them can be expressed as

$$s = \frac{2}{T_d} \frac{(1 - z^{-1})}{(1 + z^{-1})}$$

$$z = \frac{1 + (T_d/2)s}{1 - (T_d/2)s}$$

$$H(z) = H_c \left(\frac{2}{T_d} \frac{(1 - z^{-1})}{(1 + z^{-1})} \right) \tag{1.52}$$

where T_d is a sampling parameter representing a numerical integration step size. If $\sigma < 0$, then $|z| < 1$ for any value of Ω, and if $\sigma > 0$, then $|z| > 1$. Thus, stable poles, of $H_c(s)$, in the left-half of the complex s-plane are mapped inside the unit circle in the complex z-plane.

After some simplification, the relation between the continuous time frequency variable, Ω, and the discrete time frequency variable, ω, as determined through the bilinear transform is

$$\Omega = \frac{2}{T_d} \tan(\omega/2)$$

$$\omega = 2 \arctan(\Omega T_d/2) \tag{1.53}$$

1.7 Summary

In this chapter we have presented the fundamental prerequisites in digital signal processing such as convolution, sampling theory, basics of linear and time-invariant (LTI) systems, minimum-phase and all-pass systems, sampling and reconstruction of signals, discrete time Fourier transform (DTFT), discrete Fourier transform (DFT), z-transform, bilinear transform, and linear-phase finite impulse response (FIR) filters.

2

Filter Design for Audio Applications

In this chapter we present a summary of various approaches for finite impulse response (FIR) and infinite duration impulse response (IIR) filter designs.

2.1 Filter Design Process

A typical filter design approach includes the following steps.

- Specify a desired response $H_d(e^{j\omega})$ (including magnitude and/or phase specification).
- Select an FIR (or IIR) model filter having frequency response $H(e^{j\omega})$ for modeling the desired response.
- Establish a weighted or unweighted (frequency domain or time domain) approximation error criterion for comparing $H_d(e^{j\omega})$ with $H(e^{j\omega})$.
- Minimize the error criterion by optimizing the model filter parameters.
- Analyze the model filter performance (error criterion, computational complexity, etc.).

2.1.1 Desired Response Specification

The desired response can be specified in the frequency domain (viz., $H_d(e^{j\omega})$) or in the time domain (e.g., $h_d(n) = \delta(n - n_d)$). For example, a low-pass filter specification is

$$H_d(e^{j\omega}) = \begin{cases} 1 & \omega \in [0, \omega_c] \\ 0 & \omega \in [\omega_s, \pi] \end{cases} \tag{2.1}$$

The domains $[0, \omega_c]$, (ω_c, ω_s), $[\omega_s, \pi]$ are called the pass-band, transition-band, and stop-band, respectively, and are specified by their tolerance parameters. Examples of tolerance parameters include allowable ripples δ_p and δ_s which describe the pass-band *amplitude* variance A_p and stop-band attenuation A_s.

$$A_p = 20 \log_{10}(1 + \delta_p)/(1 - \delta_p) \quad \text{(dB)}$$

$$A_s = -20 \log_{10}(\delta_s) \quad \text{(dB)} \tag{2.2}$$

Alternatively, if the signal waveform needs to be preserved, then the phase response of the desired response is specified with linearity constraint,

$$\phi(e^{j\omega}) = -\tau_0\omega + \tau_1 \tag{2.3}$$

where τ_0 and τ_1 are constants.

2.1.2 Approximating Error Function

The specifications of (2.1) and (2.2) of the low-pass filter, for example, can also be written in terms of a frequency weighting approximation such as

$$-\delta_p \leq W(e^{j\omega})(|H(e^{j\omega})| - H_d(e^{j\omega})) \leq \delta_p \quad \omega \in X_p$$
$$W(e^{j\omega})|H(e^{j\omega})| \leq \delta_s \quad \omega \in X_s \tag{2.4}$$

where the accuracy of the amplitude of the selected filter, $H(e^{j\omega})$, in the pass-band domain, X_p and stop-band domain, X_s, is controlled by the frequency weighting function, $W(e^{j\omega})$.

Thus, according to (2.4) the frequency weighted approximating error function, $E(e^{j\omega})$, can be written as $E(e^{j\omega}) = W(e^{j\omega})(|H(e^{j\omega})| - H_d(e^{j\omega}))$, with $H_d(e^{j\omega}) = 0$ on the stop-band domain X_s.

Other widely used error criteria are:

• Minimax error, where ϵ, the maximum error in a particular frequency band, is minimized. Specifically, $\epsilon = \max_{\omega \in X} |E(e^{j\omega})|$.

• Minimization of the L_p norm, where the minimization of the quantity, J_p. $\int_{\omega \in X} E^p(e^{j\omega}) d\omega$, is done for $p > 0$. When $p \to \infty$, the solution that minimizes the integration approaches the minimax solution. The classic case is the L_2 norm where $p = 2$.

• Maximally flat approximation, which is obtained by means of a Taylor series expansion of the desired response at a particular frequency point.

• Combination of any of the above approximating schemes.

2.2 FIR Filter Design

There are many advantages of using FIR filters (over their IIR counterparts) which include [7] linear-phase constraint design, computationally efficient realizations, stable designs free of limit cycle oscillations when implemented on finite-wordlength systems, arbitrary specification-based designs, low output noise due to multiplication roundoff errors, and low sensitivity to variations in the filter coefficients. The disadvantages include a larger length filter for extremely narrow or stringent transition bands thereby increasing the computational requirements but which can be minimized through fast convolution algorithms and multiplier-efficient realizations.

2.2.1 Linear Phase Filter Design

There are four types of causal linear phase responses, of finite duration or finite impulse response (FIR).

1. Type 1 linear phase filter of length $M + 1$ (M even, constant group delay $M/2, \beta = \{0, \pi\}$) having finite duration response $h(n) = h(M - n)$, and frequency response $H(e^{j\omega}) = e^{(-j\omega M/2)} \sum_{k=0}^{M/2} a_k \cos(k\omega)$ with $a_0 = h(M/2)$ and $a_k = 2h((M/2) - k), 1 \leq k \leq M/2$. Type 1 filters are used to design low-pass, high-pass, and band-pass filters.

2. Type 2 linear phase filter of length $M + 1$ (M odd, a delay $M/2$ corresponding to an integer plus one-half, $\beta = \{0, \pi\}$) having finite duration response $h(n) = h(M - n)$, and frequency response $H(e^{j\omega}) = e^{(-j\omega M/2)} \sum_{k=1}^{(M+1)/2} b_k \cos(\omega(k - (1/2)))$ with $b_k = 2h((M + 1)/2 - k), 1 \leq k \leq (M + 1)/2$. Type 2 filters have a zero at $z = -1$ (i.e., $\omega = \pi$) and hence cannot be used for designing high-pass filters.

3. Type 3 linear phase filter of length $M + 1$ (M even, a delay $M/2$, $\beta = \{\pi/2, 3\pi/2\}$) having finite duration response $h(n) = -h(M - n)$, and frequency response $H(e^{j\omega}) = je^{(-j\omega M/2)} \sum_{k=1}^{M/2} c_k \sin(k\omega)$ with $c_k = 2h((M/2) - k)$, $1 \leq k \leq M/2$. Type 3 filter has a zero at $z = 1$ and $z = -1$ and hence cannot be used for designing a low-pass or a high-pass filter.

4. Type 4 linear phase filter of length $M + 1$ (M odd, $M/2$ being an integer plus one-half, $\beta = \{\pi/2, 3\pi/2\}$) having finite duration response $h(n) = -h(M - n)$, and frequency response $H(e^{j\omega}) = je^{(-j\omega M/2)} \sum_{k=1}^{(M+1)/2} d_k \sin(\omega(k - (1/2))$ with $d_k = 2h((M + 1)/2 - k), 1 \leq k \leq (M + 1)/2$. Type 4 filter has a zero at $z = 1$ and hence cannot be used in the design of a low-pass filter.

For simplicity in notation in subsequent sections, the general linear-phase filter frequency response can be described in the following functional form,

$$H(e^{j\omega}) = \sum_n t_n \psi(\omega, n) \tag{2.5}$$

where the trigonometric function $\psi(\cdot, \cdot)$ is a symbolic description for the sin or cos term in the four types of linear phase filters described above.

The design of linear-phase FIR filters, depending on the zero locations of these filters, is shown in [2]. As in the case of the decomposition of a general transfer function into minimum-phase and all-pass components, any linear-phase system function can also be decomposed into a product of three terms comprising: (i) a minimum-phase function, (ii) a maximum-phase function (where all of the poles and zeros have magnitude strictly greater than unity), and (iii) a function comprising zeros having strictly unit magnitude.

2.2.2 Least Squares FIR Filter Design

The least squares FIR filter design approximation criteria is given as

$$J_2 = \int_{\omega \in X} E^2(e^{j\omega}) d\omega = \int_{\omega \in X} [W(e^{j\omega})(|H(e^{j\omega})| - H_d(e^{j\omega}))]^2 d\omega \quad (2.6)$$

For a discrete frequency representation $\{\omega_k : k = 1, \ldots, K\}$, (2.6) can be recast as

$$J_2 = \sum_{k=1}^{K} [W(e^{j\omega_k})(|H(e^{j\omega_k})| - H_d(e^{j\omega_k}))]^2 \quad (2.7)$$

In order to design a linear-phase FIR filter model to approximate $H_d(e^{j\omega_k}) \forall k$, as given in the previous section, the generalized functional representation, $H(e^{j\omega_k}) = \sum_{n=0}^{M} t_n \psi(\omega_k, n)$, is used. Thus,

$$J_2 = \sum_{k=1}^{K} \left[W(e^{j\omega_k}) \left(\sum_{n=0}^{M} t_n \psi(\omega_k, n) - H_d(e^{j\omega_k}) \right) \right]^2 \quad (2.8)$$

which can be expressed in matrix-vector notation as,

$$J_2 = \mathbf{e}^T \mathbf{e} \quad (2.9)$$

where

$$\mathbf{e} = \mathbf{X}\mathbf{t} - \mathbf{d} \quad (2.10)$$

where the matrix $\mathbf{X} \subset \Re^{K \times (M+1)}$ is given by

$$
\begin{aligned}
\mathbf{X} = [&W(\omega_1)\psi(\omega_1, 0) \quad W(\omega_1)\psi(\omega_1, 1) \quad \cdots \quad W(\omega_1)\psi(\omega_1, M); \\
&W(\omega_2)\psi(\omega_2, 0) \quad W(\omega_2)\psi(\omega_2, 1) \quad \cdots \quad W(\omega_2)\psi(\omega_2, M); \\
&W(\omega_K)\psi(\omega_K, 0) \quad W(\omega_K)\psi(\omega_K, 1) \quad \cdots \quad W(\omega_K)\psi(\omega_K, M)] \quad (2.11)
\end{aligned}
$$

The vectors $\mathbf{t} \subset \Re^{(M+1) \times 1}$ and $\mathbf{d} \subset \Re^{K \times 1}$ are

$$
\begin{aligned}
\mathbf{t} &= (t_0, t_1, \ldots, t_M)^T \\
\mathbf{d} &= (W(\omega_1)\psi(\omega_1, 0)H_d(e^{j\omega_1}), \ldots, W(\omega_K)\psi(\omega_K, 0)H_d(e^{j\omega_K}))^T \quad (2.12)
\end{aligned}
$$

The least squares optimal solution is then given by

$$\mathbf{t} = (\mathbf{X}^T \mathbf{X})^{-1} \mathbf{X}^T \mathbf{t} \quad (2.13)$$

2.2.3 FIR Windows for Filter Design

In many instances the FIR (or IIR) filters designed can be of very large duration which will increase the computational requirements for implementing such filters, which are typically not available in real-time DSP devices. Then one approach is to limit the duration of the filter without significantly affecting the resulting performance of the filter. There are several *windowing* filters that limit the signal duration and achieve a tradeoff between the main-lobe width and side-lobe amplitudes.

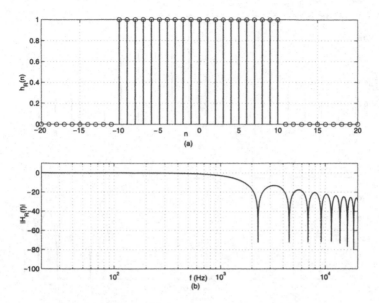

Fig. 2.1. (a) Impulse response of a rectangular window function $h_r(n)$ with $N = 10$; (b) magnitude response of the rectangular window.

A direct truncation of a signal $x(n)$ with a rectangular window filter $h_r(n)$ gives a shortened duration signal $x_r(n)$,

$$x_r(n) = h_r(n)x(n)$$
$$h_r(n) = \begin{cases} 1 & n \in \{-N, N\} \\ 0 & |n| > N \end{cases} \tag{2.14}$$

The frequency response of the rectangular window is given by $H_r(e^{j\omega}) = \sin[(2N + 1)\omega/2]/\sin(\omega/2)$. The time domain response and the magnitude response of the rectangular window are shown in Fig. 2.1.

The Bartlett window time domain and frequency response are given by $h_{Bt} = 1 - (|n|/(N + 1))$ and $H_{Bt}(e^{j\omega}) = (1/(N + 1))[\sin((N + 1)\omega/2)/(\sin(\omega/2))]^2$ and are shown in Fig. 2.2.

The Hann window is given by $h_{Hn} = 0.5(1 + \cos(2\pi n/(2N + 1)))$ and its frequency response, $H_{Hn}(e^{j\omega})$, is expressed in relation to the frequency response of the rectangular window, $H_r(e^{j\omega})$, as

$$H_{Hn}(e^{j\omega}) = 0.5H_r(e^{j\omega}) + 0.25H_r\left(e^{j(\omega - 2\pi/(2M+1))}\right)$$
$$+ 0.25H_r\left(e^{j(\omega + 2\pi/(2M+1))}\right).$$

The time domain and magnitude response for this window are shown in Fig. 2.3.

The Hamming window is given by $h_{Hm} = 0.54(1 + 0.8519\cos(2\pi n/(2N+1)))$ and its frequency response, $H_{Hm}(e^{j\omega})$, is expressed in relation to the frequency

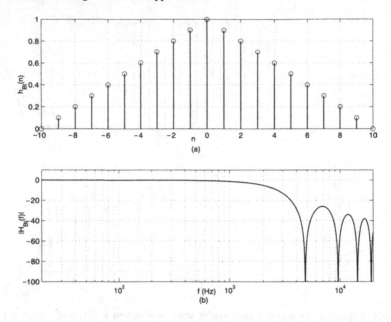

Fig. 2.2. (a) Impulse response of a Bartlett window function with $N = 10$; (b) magnitude response of the Bartlett window.

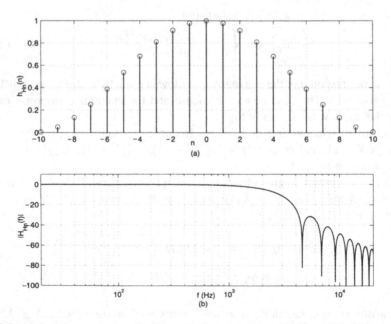

Fig. 2.3. (a) Impulse response of a Hann window function with $N = 10$; (b) magnitude response of the Hann window.

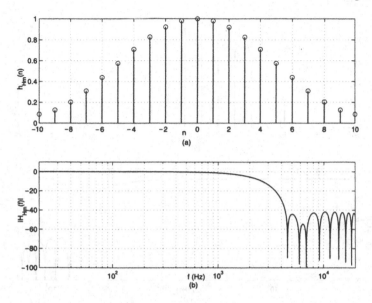

Fig. 2.4. (a) Impulse response of a Hamming window function with $N = 10$; (b) magnitude response of the Hamming window.

response of the rectangular window, $H_r(e^{j\omega})$, as $H_{Hm}(e^{j\omega}) = 0.54H_r(e^{j\omega}) + 0.23H_r(e^{j(\omega-2\pi/(2M+1))}) + 0.23H_r(e^{j(\omega+2\pi/(2M+1))})$. The time domain and magnitude response for this window are shown in Fig. 2.4.

The Blackman window is given by $h_{Bl} = 0.42 + 0.5\cos(2\pi n/(2N + 1)) + 0.08\cos(4\pi n/(2N + 1))$ and its frequency response, $H_{Bl}(e^{j\omega})$, is expressed in relation to the frequency response of the rectangular window, $H_r(e^{j\omega})$, as

$$H_{Bl}(e^{j\omega}) = 0.42H_r(e^{j\omega}) + 0.25H_r\left(e^{j(\omega-2\pi/(2M+1))}\right)$$
$$+ 0.25H_r\left(e^{j(\omega+2\pi/(2M+1))}\right) + 0.04H_r\left(e^{j(\omega-4\pi/(2M+1))}\right)$$
$$+ 0.04H_r\left(e^{j(\omega+4\pi/(2M+1))}\right).$$

The time domain and magnitude response for this window are shown in Fig. 2.5.

The Kaiser window is given by $h_K = I_0[\beta(1 - [(n - \alpha)/\alpha]^2)^{0.5}]/I_0(\beta), 0 \leq n \leq N$ where I_0 represents the zeroth-order modified Bessel function of the first kind. The main lobe width and side-lobe levels can be adjusted by varying the length $(N + 1)$ and β. The parameter, β, performs a tapering operation with high values of β achieving a sharp taper. In the extreme case, where $\beta = 0$, the Kaiser window becomes the rectangular window $h_R(n)$. Figure 2.6(a) shows the Kaiser window response for various values of β, whereas Figure 2.6(b) shows the corresponding magnitude response which shows a lower side-lobe level but increasing main-lobe width as β increases in value. Fig. 2.7 shows the magnitude responses of the Kaiser window filter, as a function of the filter or window length, N, with $\beta = 6$. As is

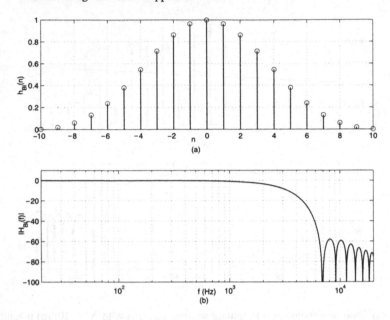

Fig. 2.5. (a) Impulse response of a Blackman window function with $N = 10$; (b) magnitude response of the Blackman window.

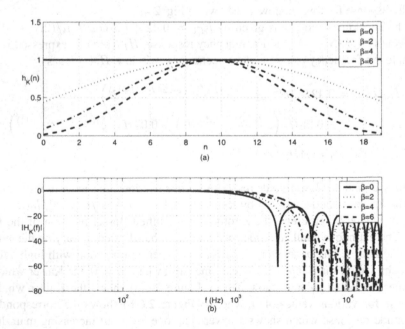

Fig. 2.6. (a) The effect of β on the shape of the Kaiser window; (b) the magnitude response of the Kaiser window for the various values of β.

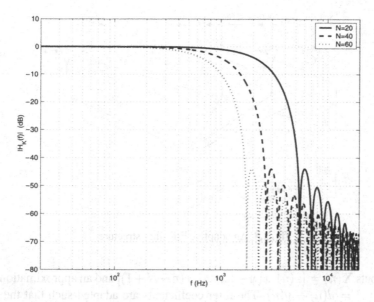

Fig. 2.7. The effect of N on the magnitude response of the Kaiser window with $\beta = 6$.

evident as the filter length increases, advantageously the main-lobe width as well as the side-lobe amplitude decreases. Kaiser determined empirically that to achieve a specified stop-band ripple amplitude $A = -20 \log_{10} \delta_s$, the value of β can be set as

$$\beta = \begin{cases} 0.1102(A - 8.7) & A > 50 \\ 0.5842(A - 21)^{0.4} + 0.07886(A - 21) & 21 \leq A \leq 50 \\ 0 & A < 21 \end{cases} \quad (2.15)$$

with $N = (A - 8)/(2.285\Delta\omega)$ and $\Delta\omega = \omega_s - \omega_p$, where ω_s and ω_p are the stop-band and pass-band cutoff frequencies.

Other approaches to "static" FIR filter design (e.g., the Remez exchange algorithm based on the alternation theorem) can be found in [7, 2].

2.2.4 Adaptive FIR Filters

Adaptive filtering is found widely in applications involving radar, sonar, acoustic signal processing (noise and echo cancellation, source localization, etc.), speech compression, and others. The advantage of using adaptive filters is their ability to track information from a nonstationary environment in real-time by optimization of its internal parameters. Haykin [8] provides a number of references to various types of adaptive filters and their applications. Of them, the most popular is the least mean square (LMS) algorithm of Widrow and Hoff [13] which is explained in this section.

An adaptive FIR filter (also called the transversal or tapped delay line filter) structure, shown in Fig. 2.8, differs from a fixed FIR filter in that the filter coefficients $\mathbf{W_k} = (w_0(k), w_1(k), \ldots, w_{N-1}(k))^T$ are varied with time as a function of the

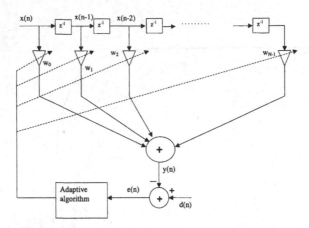

Fig. 2.8. An adaptive FIR filter structure.

filter inputs $\mathbf{X}(n) = [x(n), x(n-1), \dots, x(n-N+1)]$ and an approximation error signal $e(n) = d(n) - y(n)$. The filter coefficients are adapted such that the mean square error $J_m(n) = E\{e(n)^2\} \approx 1/m \sum_{k=0}^{m-1} e^2(n-k)$ (the operator $E\{\cdot\}$ is the statistical expectation operator) is minimized. For the well-known LMS method of [13], the instantaneous error, $J_1(n)$, where $m = 1$, is minimized.

The LMS filter adaptation equations, for a complex input signal vector $\mathbf{X}(n) = [x(n), x(n-1), \dots, x(n-N+1)]$, are expressed as

$$\mathbf{W}(n) = \mathbf{W}(n-1) + \mu e(n-1)\mathbf{X}^*(n-1) \tag{2.16}$$

where μ is the adaptation rate that controls the rate of convergence to the solution, and the superscript $*$ denotes complex conjugation. Details on the convergence and steady-state performance of adaptive filters (e.g., based on LMS, the recursive least squares or RLS error criteria) can be found in various texts and articles including [8, 126]. Other variations include the filtered-X, frequency domain, and block adaptive filters.

2.3 IIR Filter Design

The infinite duration impulse response (IIR) digital filter, $H(z)$, of numerator order and denominator order M and N, respectively, has a transfer function that resembles

$$H(z) = \frac{b_0 + b_1 z^{-1} + b_2 z^{-2} + \cdots + b_{M-1} z^{-(M-1)}}{a_0 + a_1 z^{-1} + a_2 z^{-2} + \cdots + a_{M-1} z^{-(M-1)}}$$

$$h(n) = \frac{1}{a_0} \left(-\sum_{k=1}^{N-1} a_k h(n-k) + \sum_{k=0}^{M-1} b_k x(n-k) \right) \tag{2.17}$$

Generally IIR filters can approximate specific frequency responses with shorter orders than an FIR (especially where sharp and narrow transition bands are required), but have associated problems, including (i) converging to a stable design, (ii) higher computational complexity requiring nonlinear optimization to converge to a valid solution, and (iii) numerical problems in computing an equivalent polynomial, that defines the transfer function, if multiple-order poles are densely packed near the unit circle. Furthermore, IIR filters cannot be designed to have linear phase, like FIR filters, but by cascading an all-pass filter the phase can be approximately linearized in a particular band of the frequency response.

2.3.1 All-Pass Filters

A well-known class of IIR filters, introduced in Chapter 1, are the all-pass filters having a transfer function of the form

$$H_{ap}(z) = \frac{b_0 + b_1 z^{-1} + b_2 z^{-2} + \cdots + b_{M-1} z^{-(M-1)}}{b_{M-1} + b_{M-2} z^{-1} + \cdots + b_1 z^{-(M-2)} + b_0 z^{-(M-1)}} \qquad (2.18)$$

where the coefficients of the numerator and denominator polynomial are reversed in relation to each other.

2.3.2 Butterworth Filters

Butterworth (IIR) filters are widely used in audio systems as high-pass and low-pass bass management filters. Bass management ensures that specific loudspeakers reproduce audio content, with minimal distortion, when supplied with signals in specific frequency bands.[1] The magnitude responses of the high-pass Butterworth IIR filter employed in the satellite channel, and the low-pass Butterworth IIR filter employed in the subwoofer channel can be expressed as

$$|H^{hp}_{\omega_c,N}(e^{j\omega})| = 1 - 1/\sqrt{1 + (\omega/\omega_c)^{2N}}$$

$$|H^{lp}_{\omega_c,M}(e^{j\omega})| = 1/\sqrt{1 + (\omega/\omega_c)^{2M}} \qquad (2.19)$$

The filters are second-order with a decay rate of $6N$ dB/octave and $6M$ dB/octave for the high-pass and low-pass filters, below and above the crossover frequency, f_c, respectively. For example, a typical choice of the bass management filter parameters, used in consumer electronics applications, involves $N = 2$ and $M = 4$ with a crossover frequency ω_c corresponding to 80 Hz. The magnitude responses of the bass management filters, as well as the magnitude of the recombined response (i.e., the magnitude of the complex sum of the bass management filter frequency responses), for the 80 Hz crossover with $N = 2$ and $M = 4$, are shown in Fig. 2.9.[2]

---

[1] As an example, the satellite speakers may be driven with signals above 80 Hz and the subwoofer may be driven with signals below 80 Hz.

[2] Optionally two second-order Butterworth low-pass filters (i.e., $M = 2$) are cascaded in a manner such that the speaker roll-off is initially effected with the first second-order Butterworth filter and the bass-management system includes a second second-order Butterworth filter such that the net response, from from the fourth order low-pass Butterworth and the two second-order Butterworth high-pass filters, has unit amplitude through half the sampling rate.

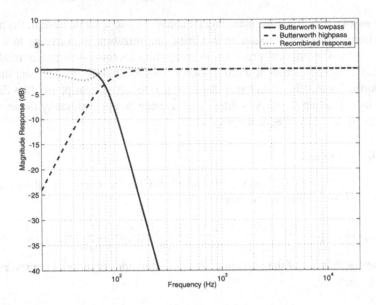

Fig. 2.9. Magnitude response of typically used bass management filters in consumer electronics, and the recombined response (the sum of the low-pass and high-pass frequency responses).

2.3.3 Chebyshev Filters

Type-1 Chebyshev IIR filters exhibit equiripple error in the pass-band and monotonically decreasing response in the stop-band. A low-pass Nth-order Chebyshev IIR filter is specified by the squared magnitude response, $H_{cheby}(e^{j\omega})$, where the magnitude response oscillates between 1 and $1/(1 + \epsilon^2)$ in the pass-band, where it will have a total of N local maxima and local minima.

$$|H_{cheby,1}(e^{j\omega})|^2 = \frac{1}{(1 + \epsilon^2 T_N^2(\omega/\omega_P))} \tag{2.20}$$

where $T_N(x)$ is the Nth-order Chebyshev polynomial. For nonnegative integers k, the kth-order Chebyshev polynomial is expressed as

$$T_k(x) = \begin{cases} \cos(k \cos^{-1} x) & |x| \leq 1 \\ \cosh(k \cosh^{-1} x) & |x| \geq 1 \end{cases} \tag{2.21}$$

A Type-1 low-pass Chebyshev filter, with pass-band frequency of 1 kHz, stop-band frequency of 2 kHz, and attenuation of 60 dB, is shown in Fig. 2.10.

Type-2 Chebyshev IIR filters exhibit equiripple error in the stop-band and the response decays monotonically in the pass-band. The squared magnitude response is expressed by,

$$|H_{cheby,2}(e^{j\omega})|^2 = \frac{1}{(1 + \epsilon^2 [T_N(\omega_s/\omega_P)]/T_N(\omega_s/\omega)]^2)} \tag{2.22}$$

and is shown in Fig. 2.11.

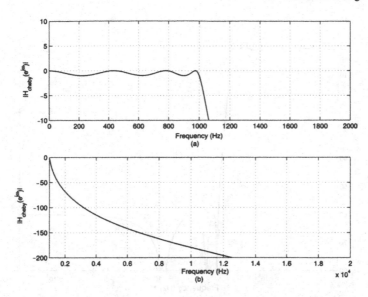

Fig. 2.10. (a) Magnitude response, between 20 Hz and 1500 Hz, of Type-1 Chebyshev low-pass filter of order 7 having pass-band frequency of 1000 Hz, stop-band frequency of 2000 Hz, and attenuation of 60 dB; (b) magnitude response of the filter in (a), between 1000 Hz and 20,000 Hz.

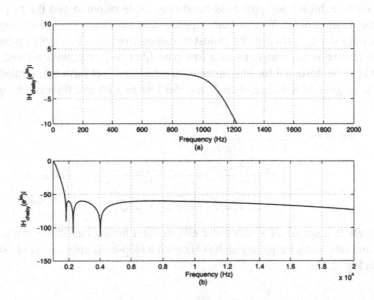

Fig. 2.11. (a) Magnitude response, between 20 Hz and 1500 Hz, of Type-2 Chebyshev low-pass filter of order 7 having pass-band frequency of 1000 Hz, stop-band frequency of 2000 Hz, and attenuation of 60 dB; (b) magnitude response of the filter in (a), between 1000 Hz and 20,000 Hz.

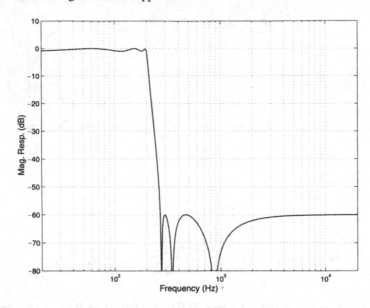

Fig. 2.12. Magnitude response of an $N = 5$-order elliptic filter having pass-band frequency of 200 Hz and stop-band frequency of 300 Hz, with stop-band attenuation of 60 dB.

2.3.4 Elliptic Filters

Elliptic filters exhibit equiripple pass-band magnitude response and the stop-band. For a specific filter order N, pass-band ripple ϵ, and maximum stop-band amplitude $1/A$, the elliptic filter provides the fastest transition from pass-band to stop-band. In fact, this feature is advantageous as a low-pass filter response, with a rapid decay time, and can be designed for low-order transversal or direct form two implementations. The magnitude response of an Nth-order low-pass elliptic filter can be written as

$$|H(e^{j\omega})|^2 = \frac{1}{1 + \epsilon^2 F_N^2(\omega/\omega_P)}$$

$$F_N(\omega) = \begin{cases} \gamma^2 \frac{(\omega_1^2-\omega^2)(\omega_3^2-\omega^2)...(\omega_{2N-1}^2-\omega^2)}{(1-\omega_1^2\omega^2)(1-\omega_3^2\omega^2)...(1-\omega_{2N-1}^2\omega^2)} & N, \text{even} \\ \gamma^2 \frac{\omega(\omega_2^2-\omega^2)(\omega_4^2-\omega^2)...(\omega_{2N}^2-\omega^2)}{(1-\omega_2^2\omega^2)(1-\omega_4^2\omega^2)...(1-\omega_{2N}^2\omega^2)} & N, \text{odd} \end{cases} \quad (2.23)$$

The magnitude response of a fifth-order elliptic filter having pass-band frequency at 200 Hz and stop-band frequency at 300 Hz with a stop-band attenuation of 60 dB is shown in Fig. 2.12.

2.3.5 Shelving and Parametric Filters

Commonly used IIR filters in audio applications include the second-order parametric filter for designing filters with specific gain and bandwidth (or Q value) and the shelving filter which introduces amplitude boosts or cuts in low-frequency or high-frequency regions. A second-order transfer function is expressed as

$$H(z) = \frac{b_0 + b_1 z^{-1} + b_2 z^{-2}}{a_0 + a_1 z^{-1} + a_2 z^{-2}} \qquad (2.24)$$

and the coefficients a_i and b_j for the various filters are given below.

Low-Frequency Shelving Filter

1. Boost of G (dB) (i.e., $g = 10^{G/20}$) with $K = \tan(\Omega_c T_s / 2)$:

$$b_0 = \frac{1 + \sqrt{2g}K + gK^2}{1 + \sqrt{2}K + K^2}$$

$$b_1 = \frac{2(gK^2 - 1)}{1 + \sqrt{2}K + K^2}$$

$$b_2 = \frac{1 - \sqrt{2g}K + gK^2}{1 + \sqrt{2}K + K^2}$$

$$a_0 = 1 \qquad (2.25)$$

$$a_1 = \frac{2(K^2 - 1)}{1 + \sqrt{2}K + K^2}$$

$$a_2 = \frac{1 - \sqrt{2}K + K^2}{1 + \sqrt{2}K + K^2}$$

2. Cut of G (dB) (i.e., $g = 10^{G/20}$) with $K = \tan(\Omega_c T_s / 2)$:

$$b_0 = \frac{1 + \sqrt{2}K + K^2}{1 + \sqrt{2g}K + gK^2}$$

$$b_1 = \frac{2(K^2 - 1)}{1 + \sqrt{2g}K + gK^2}$$

$$b_2 = \frac{1 - \sqrt{2}K + K^2}{1 + \sqrt{2g}K + gK^2}$$

$$a_0 = 1 \qquad (2.26)$$

$$a_1 = \frac{2(gK^2 - 1)}{1 + \sqrt{2g}K + gK^2}$$

$$a_2 = \frac{1 - \sqrt{2g}K + gK^2}{1 + \sqrt{2g}K + gK^2}$$

Figures 2.13 and 2.14 show examples of the magnitude response for a low-frequency boost and cut shelving filters, respectively, for a 48 kHz sampling rate with $f_c = 200$ Hz or $\Omega_c = 400\pi$ and $G = 10$ dB.

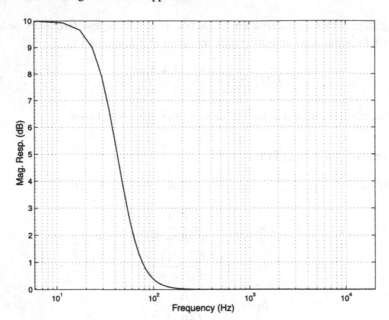

Fig. 2.13. Magnitude response for a low-frequency boost shelving filter.

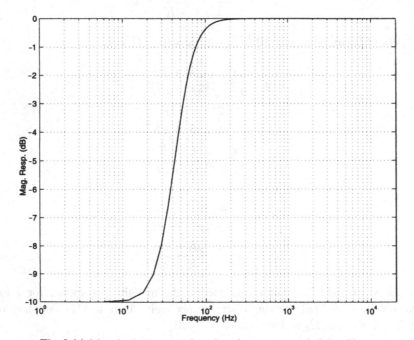

Fig. 2.14. Magnitude response for a low-frequency cut shelving filter.

High-Frequency Shelving Filter

1. Boost of G (dB) (i.e., $g = 10^{G/20}$) with $K = \tan(\Omega_c T_s/2)$:

$$b_0 = \frac{g + \sqrt{2g}K + K^2}{1 + \sqrt{2}K + K^2}$$

$$b_1 = \frac{2(K^2 - g)}{1 + \sqrt{2}K + K^2}$$

$$b_2 = \frac{g - \sqrt{2g}K + K^2}{1 + \sqrt{2}K + K^2}$$

$$a_0 = 1 \qquad\qquad (2.27)$$

$$a_1 = \frac{2(K^2 - 1)}{1 + \sqrt{2}K + K^2}$$

$$a_2 = \frac{1 - \sqrt{2}K + K^2}{1 + \sqrt{2}K + K^2}$$

2. Cut of G (dB) (i.e., $g = 10^{-G/20}$) with $K = \tan(\Omega_c T_s/2)$:

$$b_0 = \frac{1 + \sqrt{2}K + K^2}{g + \sqrt{2g}K + K^2}$$

$$b_1 = \frac{2(K^2 - 1)}{g + \sqrt{2g}K + K^2}$$

$$b_2 = \frac{1 - \sqrt{2}K + K^2}{g + \sqrt{2g}K + K^2}$$

$$a_0 = 1 \qquad\qquad (2.28)$$

$$a_1 = \frac{2((K^2/g) - 1)}{1 + \sqrt{2/g}K + K^2/g}$$

$$a_2 = \frac{1 - \sqrt{2/g}K + K^2/g}{1 + \sqrt{2/g}K + K^2/g}$$

Parametric Filters

Parametric filters are specified in terms of the gain, G ($g = 10^{G/20}$), center frequency f_c, and the Q value which is inversely related to the bandwidth of the filter. The equations characterizing the second-order parametric filter for a sampling frequency of f_s are

$$\omega_c = 2\pi f_c/f_s$$
$$\beta = (2\omega_c/Q) + \omega_c^2 + 4$$
$$b_0 = [(2g\omega_c/Q) + \omega_c^2 + 4]/\beta$$
$$b_1 = (2\omega_c^2 - 8)/\beta$$

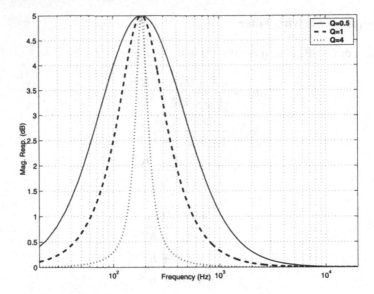

Fig. 2.15. Magnitude response of a parametric filter with various Q values.

$$b_2 = [4 - (2g\omega_c/Q) + \omega_c^2]/\beta$$
$$a_0 = 1 \qquad\qquad (2.29)$$
$$a_1 = (2\omega_c^2 - 8)/\beta$$
$$a_2 = [4 - (2\omega_c/Q) + \omega_c^2]/\beta$$

Figure 2.15 shows the effect of varying Q from 0.5 through 4 on the magnitude response of the parametric filter with $f_{center} = 187$ Hz and $G = 5$ dB.

2.3.6 Autoregressive or All-Pole Filters

IIR filters based on the autoregressive (AR) or autoregressive and moving average (ARMA) process are determined based on the second-order statistics of the input data. These filters are widely used for spectral modeling and the denominator polynomial for the AR process (or the numerator and denominator polynomials for the ARMA process) are generated through an optimization process that minimizes an error norm.

One of the popular AR processes, yielding an all-pole IIR filter, is the linear predictive coding (LPC) filter or model. The LPC model is widely used in speech recognition applications [10]: (i) it provides an excellent all-pole vocal tract spectral envelope model for a speech signal; (ii) the filter is minimum-phase, is analytically tractable, and straightforward to implement in software or hardware; and (iii) the model works well in speech recognition applications.

The LPC or all-pole filter of order P is characterized by the polynomial coefficients $\{a_k, k = 1, \ldots, P\}$ with $a_0 = 1$. Specifically, a signal $x(n)$ at time index n can be modeled with an all-pole filter of the form

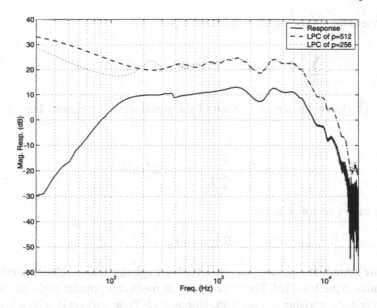

Fig. 2.16. Modeling performance of the LPC for differing filter orders.

$$H(z) = \frac{1}{1 + \sum_{k=1}^{P} a_k z^{-k}} \tag{2.30}$$

In order to determine the filter coefficients $\{a_k, k = 1, \ldots, P\}$, an intermediate error signal is defined as $e'(n) = x(n) - h(n)$. Thus,

$$E'(z) = X(z) - H(z)$$

$$= X(z) - \frac{1}{1 + \sum_{k=1}^{P} a_k z^{-k}} \left(1 + \sum_{k=1}^{P} a_k z^{-k}\right) E'(z) + 1$$

$$= X(z) \left(1 + \sum_{k=1}^{P} a_k z^{-k}\right) \tag{2.31}$$

Defining a new error signal $e(n) = \sum_{k=0}^{P} a_k e(n - k) + 1$, (2.31) may be written as

$$e(n) = x(n) + \sum_{k=1}^{P} a_k x(n - k) \tag{2.32}$$

To determine $\{a_k, k = 1, \ldots, P\}$, the error signal power, $E = \sum_{n=0}^{N-1} |e(n)|^2$, is minimized with respect to the filter coefficients with N being the duration of $x(n)$. Thus, the filter coefficients are to be determined by setting the gradient of the error function E to be zero. Thus,

$$\frac{\partial E}{\partial a_k} = 0; \quad \forall k \tag{2.33}$$

giving rise to the following all-pole *normal equations*,

$$\sum_{l=1}^{P} a_l r_x(k,l) = -r_x(k,0); \quad k = 1, 2, \ldots, P \tag{2.34}$$

where $r_x(k,l)$ denotes the correlation of the signal $x(n)$ for various lags. Specifically,

$$r_x(k,l) = \sum_{n=0}^{N-1} x(n-l)x^*(n-k) \tag{2.35}$$

or using matrix vector notation,

$$\mathbf{R}_x \mathbf{a} = -\mathbf{r}_x \tag{2.36}$$

The above system of equations can be solved through the autocorrelation method or the covariance method [10]. The autocorrelation method is popular as the autocorrelation matrix, comprising the autocorrelations $r_x(k,l)$, at various lags, is a Toeplitz matrix (i.e., a symmetric matrix with equal diagonal elements). Such a system can be solved through well-established processes such as the Durbin algorithm.

An example of the modeling performance on using the LPC approach is shown in Fig. 2.16 where the solid line depicts the response to be modeled by the LPC. The order of the LPC is $p = 128$ and $p = 256$, and it can be seen that the model shows a very good approximation at higher frequencies. Unfortunately, for such low filter orders, necessary for real-time implementations, the low-frequency performance is not satisfactory. In Chapter 6, we present a technique widely used for improving the low-frequency modeling performance of the LPC algorithm.

2.4 Summary

In this chapter we have presented various filter design techniques including FIR, IIR, parametric, shelving, and all-pole filters using second-order statistical information for signal modeling.

Part II

Acoustics and Auditory Perception

Part B

Acoustics and Auditory Perception

3

Introduction to Acoustics and Auditory Perception

This chapter introduces the theory behind sound propagation in enclosed environments, room acoustics, reverberation time, and the decibel scale. Also included are basics of loudspeakers and microphone acoustics and responses, room impulse responses, and stimuli for measuring loudspeaker and room responses. We conclude the chapter with a brief discussion on the structure of the ear, and some relevant concepts such as loudness perception and frequency selectivity.

3.1 Sound Propagation

Any complex sound field can be represented as a linear superposition of numerous simple sound waves such as plane waves. This is particularly true in the case of room acoustics, where boundaries such as walls are a source of reflections. In all of the analysis below, we assume that the medium is homogeneous and at rest, in which case the speed of sound, c, is constant with reference to space and time and is only a function of temperature, Θ. Thus,

$$c = 331.4 + 0.6\Theta \text{ m/s} \tag{3.1}$$

The spectrum of interest for a signal from a sound source, that is affected by room acoustics, is frequency dependent and is a function of the type of source. For example, human speech comprises a fundamental frequency located between 50 Hz and 350 Hz, and is identical to the frequency of vibrations of the vocal chords, in addition to harmonics that extend up to about 3500 Hz. Musical instruments range from 16 Hz to about 15 kHz. Above 10 kHz, the attenuation of the signal in air is so large that the influence of a room on high-frequency sound components can be neglected [11], whereas below 50 Hz the wavelength of sound is so large that sound propagation analysis using geometrical considerations is almost of no use. Thus, the frequency range of relevance to room acoustics extends from 50 Hz to 10 kHz.

Finally, the directionality (or the intensity of sound as a function of direction) of a sound source will vary with the type of source. For example, for speech, the directionality is less pronounced at low frequencies as the wavelengths of sound, λ (i.e., $\lambda = c/f$, where f is the frequency), at such low frequencies are less affected by diffraction effects around the head (or head shadowing), whereas musical instruments display pronounced directivity as the linear dimensions of the instrument are large compared to the wavelengths of emitted sound.

3.2 Acoustics of a Simple Source in Free-Field

The propagation of an acoustic wave in three dimensions can be modeled through a linear relationship (referred to as the *wave equation*), relating pressure p at position $\mathbf{r} = (x, y, z)$ and time t, as

$$\nabla^2 p = \frac{1}{c^2}\frac{\partial^2 p}{\partial t^2} \tag{3.2}$$

where $\nabla^2 p = (\partial^2 p/\partial x^2) + (\partial^2 p/\partial y^2) + (\partial^2 p/\partial z^2)$. The solution to the wave equation, describing the sound pressure, p, is given in terms of two arbitrary functions, $f_1(\cdot)$ and $f_2(\cdot)$, as

$$p(x, t) = f_1(ct - x) + f_2(ct + x) \tag{3.3}$$

where $f_1(\cdot)$ and $f_2(\cdot)$ describe acoustic waves traveling in the positive and negative x-direction, respectively. The functions $f_1(\cdot)$ and $f_2(\cdot)$ can be sinusoidal functions that satisfy (3.2). Specifically,

$$p(x, t) = p_0 e^{jk(ct - x)} \tag{3.4}$$

where $k = \omega/c = 2\pi/\lambda$ is the wavenumber corresponding to analog frequency ω. The plane wave assumption is valid whenever, at a distance r from the source, $kr >> 1$. Furthermore, to accommodate the nonvanishing dimensions of real-world sound sources (viz., loudspeakers), (3.4) can be generalized as

$$p(r, \phi, \theta, t) = (A/r)\Gamma(\phi, \theta)e^{jk(ct - r)} \tag{3.5}$$

where r is the distance of the sound pressure measurement point and $\Gamma(\phi, \theta)$ represents the directivity function of the source, normalized to unity.

In a more specific form, the free-field, or unbounded medium, sound pressure p_ω at any point $\mathbf{r}$ due to a simple harmonic source having an outward flow of $S_\omega e^{-j\omega t}$ from a position $\mathbf{r}_0$ can be written as

$$p_\omega(\mathbf{r}|\mathbf{r}_0)e^{-j\omega t} = -jk\rho c S_\omega e^{-j\omega t} g_\omega(\mathbf{r}|\mathbf{r}_0)e^{-j\omega t}$$

$$g_\omega(\mathbf{r}|\mathbf{r}_0) = \frac{1}{4\pi R}e^{jkR} \tag{3.6}$$

$$R^2 = |\mathbf{r} - \mathbf{r}_0|^2 = (x - x_0)^2 + (y - y_0)^2 + (z - z_0)^2$$

where k is referred to as the wavenumber and relates to the harmonic frequency ω through $k = \omega/c = 2\pi/\lambda$, ρ is the density of air,[1] and $g_\omega(\mathbf{r}|\mathbf{r}_0)$ is called the Green's

---

[1] This is a function of temperature of the air, moisture, and barometric pressure, but for all practical purposes a value of 1.25 kg/m³ can be used.

function. The time-dependent pressure function $p(\mathbf{r}|\mathbf{r}_0, t)$ can be found through the Fourier inverse of (3.2).

Finally, the sound pressure level at a distance r, with $\tilde{p}$ representing the root mean square pressure (viz., $\sqrt{E\{p^2\}} = [(1/t) \int_t p^2 d\tau]^{1/2}$ where $E\{.\}$ is the statistical expectation operator), can be expressed as

$$SPL = 20 \log_{10} \frac{\tilde{p}}{\tilde{p}_{ref}} \ \mathrm{dB} \tag{3.7}$$

where $\tilde{p}_{ref}$ is an internationally standardized reference root mean square pressure with a value of 2×10^{-5} N/m^2.

3.3 Modal Equations for Characterizing Room Acoustics at Low Frequencies

The free-field sound propagation behavior was described in the previous section. However, the sound field in a real room is much more complicated to characterize because of the large number of reflection components, standing waves influenced by room dimensions, and the variabilities in the geometry, size, and absorption between various rooms. For example, if a wave strikes a wall, in general, a part of the incident sound energy will be absorbed and some of it will be reflected back with some phase change. The resulting wave, created by the interference of the reflected and incident wave, is called a standing wave. Given that a room includes various walls, surfaces, and furniture, the prediction of the resulting sound field is very difficult. Hence several room models have evolved that characterize the sound field through deterministic or statistical techniques. One such model is based on the *wave theory of acoustics* where the acoustic wave equation of (3.2) is solved with boundary conditions that are set up which describe, mathematically, the acoustical properties of the walls, ceiling, floor, and other surfaces.

Without going into the derivation, the Green's function, or the sound pressure in a room for a frequency ω, derived from the wave theory of acoustics in a *bounded enclosure* is given by [11] and [12]

$$p_\omega(\underline{q}_l) = jQ\omega\rho_0 \sum_{\underline{n}} \frac{p_{\underline{n}}(\underline{q}_l)p_{\underline{n}}(\underline{q}_o)}{K_{\underline{n}}(k^2 - k_{\underline{n}}^2)}$$

$$= jQ\omega\rho_0 \sum_{n_x=0}^{N_x-1} \sum_{n_y=0}^{N_y-1} \sum_{n_z=0}^{N_z-1} \frac{p_{\underline{n}}(\underline{q}_l)p_{\underline{n}}(\underline{q}_o)}{K_{\underline{n}}(k^2 - k_{\underline{n}}^2)}$$

$$\underline{n} = (n_x, n_y, n_z); \quad k = \omega/c; \quad \underline{q}_l = (x_l, y_l, z_l)$$

$$k_{\underline{n}} = \pi \left[\left(\frac{n_x}{L_x}\right)^2 + \left(\frac{n_y}{L_y}\right)^2 + \left(\frac{n_z}{L_z}\right)^2 \right]^{1/2} \int \int_V \int p_{\underline{n}}(\underline{q}_l)p_{\underline{m}}(\underline{q}_l)dV$$

$$= \begin{cases} K_{\underline{n}} & \underline{n} = \underline{m} \\ 0 & \underline{n} \neq \underline{m} \end{cases} \tag{3.8}$$

where $k_{\underline{n}}$ are referred to as the eigenvalues, and where the eigenfunctions $p_{\underline{n}}(q_l)$ can be assumed to be orthogonal to each other under certain conditions, and the point source being at q_o. The modal equations in (3.8) are valid for wavelengths, λ, where $\lambda > (1/3)\min[L_x, L_y, L_z]$ [12]. At these low frequencies, a few standing waves are excited, so that the series terms in (3.8) converge quickly.

For a rectangular enclosure with dimensions (L_x, L_y, L_z), $q_o = (0,0,0)$, the eigenfunctions $p_{\underline{n}}(q_l)$ in (3.8) are

$$p_{\underline{n}}(q_l) = \cos\left(\frac{n_x \pi x_l}{L_x}\right) \cos\left(\frac{n_y \pi y_l}{L_y}\right) \cos\left(\frac{n_z \pi z_l}{L_z}\right)$$

$$p_{\underline{n}}(q_o) = 1$$

$$K_{\underline{n}} = \int_0^{L_x} \cos^2\left(\frac{n_x \pi x_l}{L_x}\right) dx \int_0^{L_y} \cos^2\left(\frac{n_y \pi y_l}{L_y}\right) dy \int_0^{L_z} \cos^2\left(\frac{n_z \pi z_l}{L_z}\right) dz$$

$$= \frac{L_x L_y L_z}{8} = \frac{V}{8} \tag{3.9}$$

Each of the terms in the series expansion can be considered to excite a resonant frequency of about $f_n = \omega_n/2\pi = c/\lambda_n$ Hz, with a specific amplitude and phase as determined by the numerator and denominator terms of (3.8). Because the different terms in the series expansion can be considered mutually independent, the central limit theorem can be applied to the real and imaginary parts of $p_\omega(q_l)$, according to which both quantities can be considered to be random variables obeying a nearly Gaussian distribution. Thus, according to the theory of probability, $|p_\omega(q_l)|$ follows the Rayleigh distribution. If z denotes $p_\omega(q_l)$, then the probability of finding a sound pressure amplitude between z and $z + dz$ is given by

$$P(z)dz = \frac{\pi}{2}e^{-(\pi z^2/4)}z\,dz \tag{3.10}$$

Thus, in essence, the distribution of the sound pressure amplitude is independent of the type of room, volume, or its acoustical properties. The probability distribution is shown in Fig. 3.1.

The eigenfunction distribution in the $z = 0$ plane, for a room of dimension 6 m $\times$6 m $\times$6 m, and tangential mode $(n_x, n_y, n_z) = (3, 2, 0)$ is shown in Fig. 3.2.

Finally, the time domain sound pressure, $p(\mathbf{r}, t)$, can be found through the Fourier transform using

$$p(\mathbf{r}, t) = \int_{-\infty}^{\infty} p_\omega(\mathbf{r})e^{-j\omega t}\,d\omega \tag{3.11}$$

Rooms such as concert halls, theaters, and irregular-shaped rooms deviate from the wave theory assumed rectangular shape due to the presence of pillars, columns, balconies, and other irregularities. As such, the methods of wave theory cannot be readily applied as the boundary conditions are difficult to formulate.

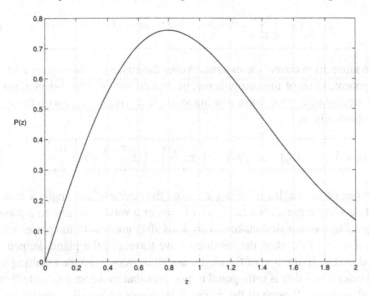

Fig. 3.1. The sound pressure amplitude density function in a room excited by a sinusoidal tone.

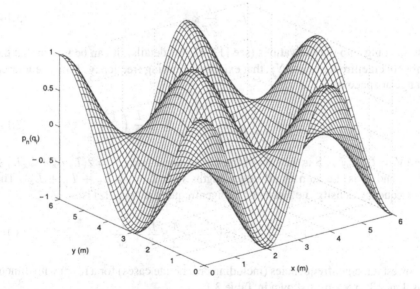

Fig. 3.2. The eigenfunction distribution, for a tangential mode (3,2,0) over a room of dimensions 6 m ×6 m ×6 m.

3.3.1 Axial, Tangential, Oblique Modes and Eigenfrequencies

By using the expression $\cos(x) = (e^{jx} + e^{-jx})/2$ in (3.9), the eigenfunction equation can be written as

$$p_{\underline{n}}(\underline{q}_l) = \frac{1}{8} \sum e^{j\pi(\pm(n_x x_l/L_x)\pm(n_y y_l/L_y)\pm(n_z z_l/L_z))} \qquad (3.12)$$

where the summation covers the expansion over the eight possible sign combinations in the exponent. Each of the components, multiplied with the time-dependent exponent $e^{j\omega t}$ represents a plane wave making an angle α_x, α_y, and α_z with the x, y, and z axis, respectively, as

$$\cos(\alpha_x) : \cos(\alpha_y) : \cos(\alpha_z) = \left(\pm \frac{n_x}{L_x}\right) : \left(\pm \frac{n_y}{L_y}\right) : \left(\pm \frac{n_z}{L_z}\right) \qquad (3.13)$$

If any one of the angles is 90 degrees (i.e., the cosine of the angle is zero), then the resulting wave represents a *tangential mode*, or a wave traveling in a plane, that is orthogonal to the axis that makes an angle of 90 degrees with the plane wave. For example, if $\alpha_z = \pi/2$, then the resulting wave travels in the plane defined by the x-axis and y-axis. If any two of the angles are 90 degrees, then the resulting wave is called an *axial mode* that is orthogonal to two axes that make an angle of 90 degrees with the plane wave. If none of the angles is 90 degrees (i.e., all of the cosine terms are nonzero) then the resulting wave represents an *oblique mode*.

The eigenfrequencies, $f_{\underline{n}}$, for the enclosure, are related to the eigenvalues, $k_{\underline{n}}$, as

$$f_{\underline{n}} = \frac{c}{2\pi} k_{\underline{n}} \qquad (3.14)$$

Without going into the derivations (see [12, 11] for details), it can be shown that the number of eigenfrequencies, N_f, that exist for a limiting frequency, f, in an enclosed rectangular space is

$$N_f = \frac{4\pi}{3} V \left(\frac{f}{c}\right)^3 + \frac{\pi}{4} S \left(\frac{f}{c}\right)^2 + \frac{L}{8} \left(\frac{f}{c}\right) \qquad (3.15)$$

where $V = L_x L_y L_z$, S is the sum of the surface areas (viz., $S = 2(L_x L_z + L_x L_y + L_y L_z)$) and L is the sum of all edge lengths with $L = 4(L_x + L_y + L_z)$. The eigenfrequency density (i.e., number of eigenfrequencies per Hz) is

$$\Delta_f = \frac{dN_f}{df} = 4\pi V \frac{f^2}{c^3} + \frac{\pi}{2} S \frac{f}{c^2} + \frac{L}{8c} \qquad (3.16)$$

The lowest ten eigenfrequencies (including degenerate cases) for a room with dimensions 4 m ×3 m ×2 m is shown in Table 3.1.

3.4 Reverberation Time of Rooms

Given a mean sound intensity $I(t)$ due to a source, such as a loudspeaker, transmitting with power $\Pi(t)$ at time t in a room of volume V and absorption of $a = \sum_i \alpha_i S_i$

$f_{\underline{n}}$ (Hz)	n_x	n_y	n_z
42.875	1	0	0
57.167	0	1	0
71.458	1	1	0
85.75	0	0	1
85.75	2	0	0
95.871	1	0	1
103.06	0	1	1
103.06	2	1	0
111.62	1	1	1
114.33	0	2	0

Table 3.1. The ten lowest eigenfrequencies for a room of dimension 4 m × 3 m × 2 m.

(where α_i and S_i are the absorption coefficient and surface area of wall i, respectively), then the rate of change of total acoustic energy in the room can be expressed through the following conservation rule,

$$\frac{d}{dt}\frac{4VI(t)}{c} = \Pi(t) - aI(t) \tag{3.17}$$

where c is the speed of sound in the medium.

The solution to (3.17) can be written as

$$I(t) = \frac{c}{4V}e^{-act/4V}\int_{-\infty}^{t}\Pi(\tau)e^{ac\tau/4V}\,d\tau \tag{3.18}$$

If the sound power $\Pi(t)$ fluctuates slowly, relative to the time constant $4V/ac$, then the intensity $I(t)$ will be approximately proportional to $\Pi(t)$ as

$$I(t) \approx \frac{\Pi(t)}{a} \tag{3.19}$$

$$\approx 10\log_{10}\frac{\Pi(t)}{a} + 90 \text{ dB} \quad \text{above} \quad 10^{-16} \text{ watt/cm}^2$$

if $\Pi(t)$ is in ergs per second and a is in square centimeters.

In the event that the sound power $\Pi(t)$ fluctuates in a time short compared to the time constant $4V/ac$, then the intensity will not follow the fluctuations of $\Pi(t)$, and if the sound is shut off suddenly at time $t = 0$, the subsequent intensity can be expressed using (3.17) as

$$I(t) = I_0 e^{-(act/4V)} \tag{3.20}$$

$$Intensity\ Level = 10\log_{10}I_0 + 90 - 4.34\frac{act}{4V} \text{ (dB)}$$

Thus, upon turning off the source, the intensity level drops off linearly at a rate of $4.34act/4V$ every dB.

The *reverberation time* of the room, which characterizes the time where the energy of reflections arriving from walls or boundary surfaces is non-negligible, is defined as the time it takes for the intensity level to drop by 60 dB after the source is switched off. Thus, if the dimensions of the room are measured in centimeters, then the reverberation time T_{60} is given by

$$T_{60} = 60\frac{4V}{4.34ac} = 0.161\frac{V}{\sum_i \alpha_i S_i} \qquad (3.21)$$

The reverberation time computed through (3.21) is based on geometrical room acoustics where the walls are considered to be sufficiently irregular so that sound energy distribution is uniform "throughout" the room. In other words the square sound pressure amplitude is independent of the distance "R" between the source and microphone and angles (α, θ) corresponding to an assemblage of plane waves reflecting from walls. Morse and Ingard [12] state that if the sound is not uniformly distributed in the room, (3.21) will not be valid and "... in fact, the term absorption coefficient will have no meaning."

The actual measurement of T_{60} can be done by the "method of integrated impulse response" proposed by Schroeder [14]. The method uses the following integration rule to determine an ensemble average of *decay curves*,$\langle g^2(t)\rangle$, from the *room impulse response*, $h(t)$,[2] using

$$\langle g^2(t)\rangle = \int_t^\infty h^2(x)dx \qquad (3.22)$$

Subsequently, the result from (3.22) is converted to dB scale and the following expression is used for computing T_{60},

$$T_{60} = 60\left(\frac{\Delta L}{\Delta t}\right)^{-1} \qquad (3.23)$$

where $\Delta L/\Delta t$ is in dB/seconds. Frequently the slope of the decay curve is determined in the range of -5 dB and -35 dB relative to the steady-state level. In addition, the time domain for integration of (3.22), in practice, is important. Of course, an upper limit of integration ∞ is not possible in real-world applications, so a finite integration interval is chosen. Care should be taken to ensure that the integration interval is not too long as the decay curve will have a tail which limits the useful dynamic range, nor should it be too short as it would cause a downward bend of the curve.

Figure 3.3 shows a room impulse response recorded in a room using a loudspeaker A, whereas the measured T_{60}, based on Fig. 3.4 (and using the measured length $L = 8192$ samples of the room response as the upper limit for integration), is found to be approximately 0.25 seconds. The effect of the upper limit of integration

---

[2] Recall from Section 1.1.4 the impulse response is the output of a linear system when the input is $\delta(n)$. In reality, room responses are applying a broadband signal to the room (such as a logarithmic chirp, noise type sequences, etc.) and measuring the result through a microphone. More information is provided in a subsequent section.

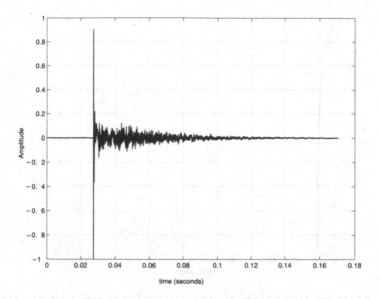

Fig. 3.3. A room impulse response measured in a room.

corresponding to $0.0625L$ is shown in Fig. 3.5, whereas the upper limit of integration is $0.5L$ in Fig. 3.6.

Figure 3.7 shows a room impulse response recorded in a room using a loud-speaker B, and at a different position, whereas the measured T_{60}, based on Fig. 3.8

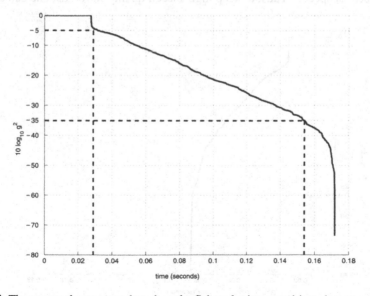

Fig. 3.4. The energy decay curve based on the Schroeder integrated impulse response technique for loudspeaker A.

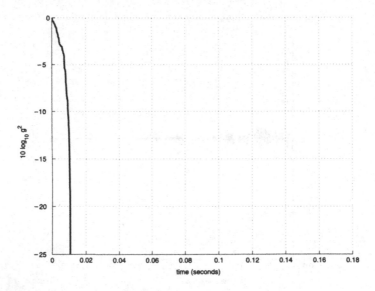

Fig. 3.5. The energy decay curve based on using $0.0625L$ as an upper limit of integration.

(and using the measured length L of the room response as the upper limit for integration), is again found to be approximately 0.25 seconds showing reasonable independence of the type of loudspeaker used in measuring the room response and the position where the response was measured.

Finally, the effect of large reverberation is that it degrades the quality of audio signals such as speech. Thus, to keep high speech quality in rooms, one can design

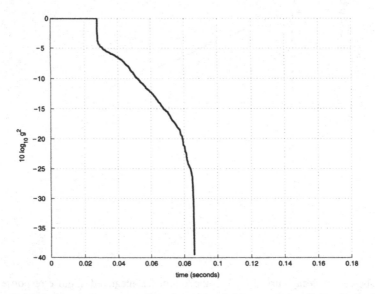

Fig. 3.6. The energy decay curve based on using $0.5L$ as an upper limit of integration.

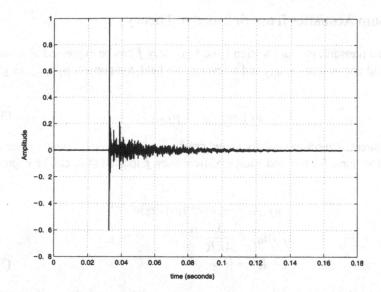

Fig. 3.7. A room impulse response measured in a room with a loudspeaker B.

the reverberation time to be small by increasing the absorption of the room a. However, this is in contradiction to the requirement that the transient intensity (3.19) be kept high. Thus, a compromise is required, during design of rooms, between these two opposing requirements.

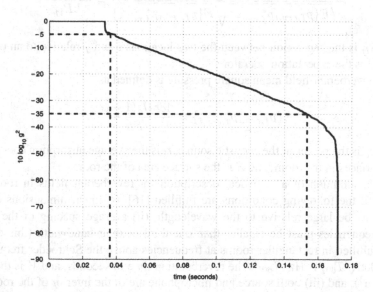

Fig. 3.8. The energy decay curve based on the Schroeder integrated impulse response technique for loudspeaker B.

3.5 Room Acoustics from Schroeder Theory

The sound pressure, $p_{f,i}$, at location i and frequency f can be expressed as a sum of direct field component, $p_{f,d,i}$, and a reverberant field component, $p_{f,rev,i}$, as given by

$$p_{f,i} = p_{f,d,i} + p_{f,rev,i} \qquad (3.24)$$

The direct field component for sound pressure, $p_{f,d,i}$, of a plane wave, at far field listener location i for a sound source of frequency f located at i_0 can be expressed as [12]

$$p_{f,d,i} = -jk\rho c S_f g_f(i|i_0) e^{-j\omega t}$$
$$g_f(i|i_0) = \frac{1}{4\pi R} e^{jkR}$$
$$R^2 = |i - i_0|^2 \qquad (3.25)$$

where $p_{f,d}(i|i_0)$ is the direct component sound pressure amplitude, S_f is the source strength, $k = 2\pi/\lambda$ is the wavenumber, $c = \lambda f$ is the speed of sound (343 m/s) and ρ is the density of the medium (1.25 kg/m^3 at sea level).

The normalized correlation function [100] which expresses a statistical relation between sound pressures, of reverberant components, at separate locations i and j, is given by

$$\frac{E\{p_{f,rev,i}p^*_{f,rev,i}\}}{\sqrt{E\{p_{f,rev,i}p^*_{f,rev,i}\}}\sqrt{E\{p_{f,rev,j}p^*_{f,rev,j}\}}} = \frac{\sin kR_{ij}}{kR_{ij}} \qquad (3.26)$$

where R_{ij} is the separation between the two locations i and j relative to an origin, and $E\{.\}$ is the expectation operator.

The reverberant-field mean square pressure is defined as

$$E\{p_{f,rev,i}p^*_{f,rev,i}\} = \frac{4c\rho \Pi_a(1 - \bar{\alpha})}{S\bar{\alpha}} \qquad (3.27)$$

where Π_a is the power of the acoustic source, $\bar{\alpha}$ is the average absorption coefficient of the surfaces in the room, and S is the surface area of the room.

The assumption of a statistical description for reverberant fields in rooms is justified if the following conditions are fulfilled [16]: (i) linear dimensions of the room must be large relative to the wavelength, (ii) average spacing of the resonance frequencies must be smaller than one-third of their bandwidth (this condition is fulfilled in rectangular rooms at frequencies above the Schroeder frequency, $f_s = 2000\sqrt{T_{60}/V}$ Hz (T_{60} is the reverberation time in seconds, and V is the volume in m^3), and (iii) both source and microphone are in the interior of the room, at least a half-wavelength away from the walls.

Furthermore, under the conditions in [16], the direct and reverberant sound pressures are uncorrelated.

3.6 Measurement of Loudspeaker and Room Responses

Measuring loudspeaker and room acoustical responses, and determining its frequency response, is one of the most important aspects in the area of acoustics and audio signal processing. In fact, even a loudspeaker designer will evaluate the loudspeaker response in an anechoic room before releasing it for production. An example of a room impulse response showing the direct path of the sound, the early reflections, and reverberation is shown in Fig. 3.9.

There are several methods for measuring speaker and/or room acoustical responses, the popular ones being based on applying a pseudo-random sequence, such as the maximum length sequence (MLS), to the loudspeaker and deconvolving the response at a microphone [17], or applying a frequency sweep signal such as the logarithmic chirp, to the speaker and deconvolving the microphone response.

Müller and Massarani [18] discuss various popular approaches for room acoustical response measurement. In this section we briefly discuss room response measurement approaches using logarithmic sweep and the maximum length sequences.

3.6.1 Room Response Measurement with Maximum Length Sequence (MLS)

The MLS-based method for finding the impulse response is based on cross-correlating a measured signal with a pseudo-random (or deterministic) sequence. The motivation for this approach is explained through the following derivation. Let $x(t)$ be a stationary sound signal having autocorrelation $\phi_{xx}(t)$, which is applied to the room with response $h(t)$ through a loudspeaker. Then the signal received at the microphone is

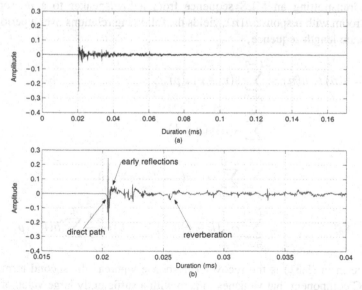

Fig. 3.9. (a) Room impulse response; (b) zoomed version of the response showing direct, early reflections, and reverberation.

$$y(t) = \int_{-\infty}^{\infty} x(t - t')h(t')dt' \tag{3.28}$$

Forming the cross-correlation, $\phi_{yx}(\tau)$ between the received signal $y(t)$ and the transmitted signal $x(t)$, we have

$$\phi_{yx}(\tau) = \lim_{T_0 \to \infty} \frac{1}{T_0} \int_{-T_0/2}^{T_0/2} \int_{-\infty}^{\infty} x(t + \tau - t')h(t')x(t)dt'\,dt$$

$$= \int_{-\infty}^{\infty} \phi_{xx}(\tau - t')h(t')dt' \tag{3.29}$$

Now if $\phi_{xx}(\tau - t') = \delta(\tau - t')^3$, then (3.29) results in the cross-correlation being equal to the room impulse response or $\phi_{yx}(t) = h(t)$.

More useful than white noise are pseudo-random signals, called MLS, which have similar properties to white noise, but are binary or two-valued in nature. Such binary sequences can be easily generated by means of a digital computer and can be processed rapidly through signal processing algorithms. Specifically, an MLS sequence, $s(n)$, of period $L = 2^n - 1$, where n is a positive integer, satisfies the following relations,

$$\sum_{k=0}^{L-1} s(k) = -1$$

$$\phi_{ss}(k) = \frac{1}{L} \sum_{n=0}^{L-1} s(n)s(n + k) = \begin{cases} 1 & k = 0, L, 2L, \dots \\ -\frac{1}{L} & k \ne 0, L, 2L, \dots \end{cases} \tag{3.30}$$

Thus, transmitting an MLS sequence from a loudspeaker to a microphone, through a room with response $h(n)$, yields the following relations over a period L of the maximum length sequence,

$$y(n) = s(n) \otimes h(n) = \sum_{p=0}^{L-1} s(n - p)h(p)$$

$$\phi_{yx}(k) = \sum_{n=0}^{L-1} s(n)y(n + k)$$

$$= \sum_{p=0}^{L-1} \sum_{n=0}^{L-1} s(n)s(n + k - p)h(p)$$

$$= \sum_{p=0}^{L-1} \phi_{ss}(k - p)h(p) = h(k) - \frac{1}{L} \sum_{p=1}^{L} h(k - p) \tag{3.31}$$

The first term in (3.31) is the recovered response whereas the second term represents a DC component that vanishes to zero with a sufficiently large value of L. An

---

[3] Signals satisfying such autocorrelation functions are referred to as white noise signals.

example of an MLS sequence with $n = 3$ or $L = 7$ is: $-1,-1,1,-1,1,-1,1$. For practical reasons, the MLS sequence is transmitted repeatedly possibly with an intervening silence interval, and the measured signal is averaged in order to improve the signal-to-noise ratio (SNR). Also, L is kept sufficiently high so as to prevent any time aliasing problems in the deconvolved response, where the late reflection (or reverberation part) folds back into the early part of the room response. This happens if the period of the repeatedly transmitted signal is smaller than the length of the impulse response $h(n)$.

Another stimulus signal constructed using the MLS signal is the inverse repeated sequence (IRS) [19], and is defined by,

$$x(n) = \begin{cases} MLS(n) & n \text{ even}, 0 \le n < 2L \\ -MLS(n) & n \text{ odd}, 0 < n < 2L \end{cases} \tag{3.32}$$

3.6.2 Room Response Measurement with Sweep Signals

Another approach for obtaining the impulse response is via a circular deconvolution approach where the measured signal is Fourier transformed, divided by the Fourier transform of the input signal, and the result inverse transformed to get the time domain impulse response. Specifically, with F and F^{-1} representing the forward and inverse Fourier transform, respectively, and $x(t)$ and $y(t)$ representing the input and measured signal,

$$h(t) = F^{-1} \left[\frac{F[y(t)]}{F[x(t)]} \right] \tag{3.33}$$

In the case of a linear sweep a constant phase increment is added to $x(t)$ (with N being the number of samples to be generated), and is given by

$$x(t) = A \cos(\phi(t))$$
$$\phi(t) = \phi(t - 1) + \Delta\phi(t)$$
$$\Delta\phi(t) = \Delta\phi(t - 1) + \psi \tag{3.34}$$
$$\psi = 2\pi \frac{f_{\text{stop}} - f_{\text{start}}}{N f_s} \tag{3.35}$$

The time domain and magnitude response plot (white excitation spectrum) for a linear sweep is shown in Fig. 3.10.

The time domain and magnitude response (3 dB per octave decay) of a logarithmic sweep [20] characterized by

$$x(t) = \sin\left(\frac{\omega_1 T}{\log \frac{\omega_2}{\omega_1}} \left(e^{(t/T) \log(\omega_2/\omega_1)} - 1 \right) \right) \tag{3.36}$$

is shown in Fig. 3.11.

The advantage of a logarithmic sweep over a linear sweep is the larger SNR at lower frequencies thereby allowing better characterization of room modes. Advantages of using logarithmic sweep over MLS is the separation of loudspeaker distortion product terms from the actual impulse response in addition to improved SNR.

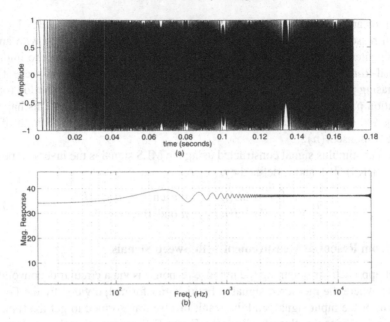

Fig. 3.10. (a) Time domain response of linear sweep; (b) magnitude response of the linear sweep.

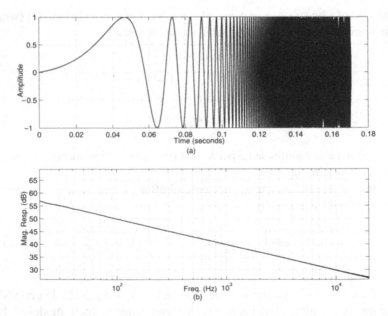

Fig. 3.11. (a) Time domain response of logarithmic sweep; (b) magnitude response of the log sweep.

Furthermore, it has been shown [21] that in the presence of nonwhite noise the MLS and IRS methods for room impulse response seem to be most accurate, whereas in quiet environments the logarithmic sine-sweep method is the most appropriate signal of choice.

3.7 Psychoacoustics

The perception of sound is an important area and recent systems employing audio compression techniques use principles from auditory perception, or psychoacoustics, for designing lower bit-rate systems without significantly sacrificing audio quality. Likewise, it seems a natural extension that certain properties of human auditory perception (e.g., frequency selectivity) be exploited to design efficient systems, such as room equalization systems, which aim at minimizing the detrimental effects of room acoustics.

3.7.1 Structure of the Ear

To understand relevant concepts from psychoacoustics, it is customary to summarize the structure of the ear. Shown in Fig. 3.12 is the peripheral part of the ear comprising the outer, middle, and inner ear sections.

The outer ear is composed of the pinna and the auditory canal. The pinna is responsible primarily for identifying the location of the source sound, particularly at high frequencies. Considerable variation exists in the conformation of the pinna and hence different people are able to localize sound differently. Sound travels down the auditory or ear canal and then strikes the tympanic membrane. The air-filled middle ear includes the tympanic membrane, the ossicles (malleus, incus, stapes), their associated muscles and ligaments, and the opening of the auditory tube, which provides communication with the pharynx as well as a route for infection. Thus,

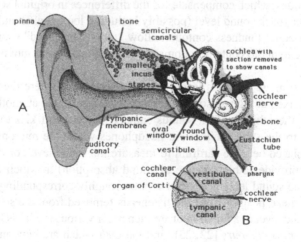

Fig. 3.12. The structure of the ear.

sound vibrations in the ear canal are transmitted to the tympanic membrane, and in turn are transmitted through the articulations of the ossicles to the attachment of the foot plate of the stapes on the membrane of the oval window. The ossicles amplify the vibrations of sound and in turn pass them on to the fluid-filled inner ear.

The cochlea, which is the snail-shaped structure, and the semicircular canals constitute the inner ear. The cochlea, enclosing three fluid-filled chambers, is encased in the temporal bone with two membranous surfaces exposed at its base (viz., the oval window and the round window). The foot plate of the stapes adheres to the oval window, transmitting sound vibrations into the cochlea. Two of the three cochlear chambers are contiguous at the apex. Inward deflections of the oval window caused by the foot plate of the stapes compress the fluid in the scala vestibuli; this compression wave travels along the coils of the cochlea in the scala vestibuli to the apex, then travels back down the coils in the scala tympani. The round window serves as a pressure-relief vent, bulging outward with inward deflections of the oval window. The third cochlear chamber, the scala media or cochlear duct, is positioned between the scala vestibuli and scala tympani. Pressure waves from sound traveling up the scala vestibuli and back down the scala tympani produce a shearing force on the hair cells of the organ of Corti in the cochlear duct. Within the cochlea, hair cell sensitivity to frequencies progresses from high frequencies at the base to low frequencies at the apex. The cells in the single row of inner hair cells passively respond to deflections of sound-induced pressure waves. Thus, space (or distance) along the cochlea is mapped to the excitation or resonant frequency, and hence the cochlea can be viewed as an auditory filtering device responsible for selective frequency amplification or attenuation depending on the frequency content of the source sound.

3.7.2 Loudness Perception

Loudness perception is an important topic as it allows design of systems that take into account sensitivity to sound intensity. For example, loudness compensation or control techniques, which compensate for the differences in original sound level and loudspeaker playback sound level (possibly including loudspeaker and room acoustics) based on equal loudness contours, allow "tonally balanced" sound perception while listening to audio content in home theater, automotive, or movie theater environments.

One way to judge loudness level is on a relative scale where the intensity of a 1 kHz tone is fixed at a given sound pressure level and a tone at another frequency is adjusted by the subject until it sounds equally loud as the 1 kHz tone. The tones are presented to the subject either via headphones where the microphone probe is inserted and placed near the eardrum to measure the sound level, or in an anechoic room (which provides a high degree of sound absorption) in which case the measurement of the sound level is done at a point roughly corresponding to the center of the listener head position after the listener is removed from the sound field. The plot of the sound level as a function of frequency for various loudness levels is called the *equal loudness contours* [22, 23], and the data which are now an International Standards Organization (ISO) standard (ISO 226, 1987)[24] are shown in Fig. 3.13.

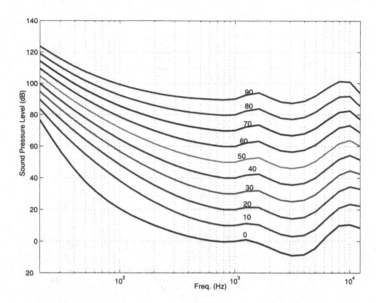

Fig. 3.13. Equal loudness contours from Robinson and Dadson [23].

The contour for 0 phon (or 0 dB SPL of a 1 kHz tone) represents the minimum audible field contour and represents on an average (among listeners) the absolute lower limit of human hearing at various frequencies. Thus, for example, at 0 phon, human hearing is most sensitive at frequencies between 3 kHz and 5 kHz as these represent the lowest part of the 0 phon curve. Furthermore, for example, the sound pressure level (SPL) has to be increased by as much as 75 dB at 20 Hz in order for a 20 Hz tone to sound equally as loud as the 1 kHz tone at 0 phon loudness level (or 0 dB SPL). In addition, the rate of growth of loudness level at lower frequencies is much greater than the middle frequencies, for example, as the SPL at 20 Hz from 0 phon to 90 phon increases by only 50 dB whereas at 1 kHz the SPL increases by 90 dB. This requirement of a larger increase in SPL at mid frequencies, relative to lower frequencies, in order to retain the same loudness level difference is the reason for the larger rate of growth of loudness level at lower frequencies. This can be observed when human voices are played back at high levels via loudspeakers, making them "boomy" as the ear becomes more sensitive to lower frequencies than higher frequencies with higher intensities.

Various SPL meters attempt giving an approximate measure for the loudness of complex tones. Such meters contain weighting networks (e.g., A, B, C, and RLB) that weigh the intensities computed in third octave frequency bands with the appropriate weighting curves before performing summation across frequencies.

The A weighting is based on a 30 phon equal loudness contour for measuring complex sounds having relatively low sound levels, the B weighting is used for intermediate levels and approximates the 70 phon contour, whereas the C weighting is for relatively high sound levels and approximates the 100 phon contour. The three weighting network contours are shown in Fig. 3.14. Thus if a level is specified as 105

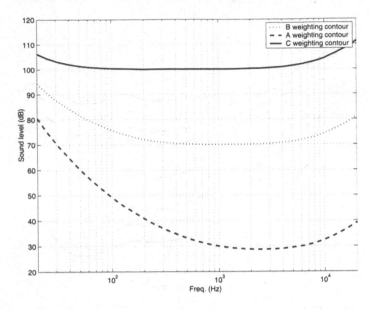

Fig. 3.14. The A, B, and C weighting networks.

dBC then the inverse C weighting (i.e., inverse of the contour shown in Fig. 3.14) is used for computing the SPL.

3.7.3 Loudness Versus Loudness Level

In the previous section the equal loudness contours were presented which were a function of the loudness level in phons. Stevens [25] presented some data that derive scales relating the physical magnitude of sounds to their subjective loudness. In this process, the subject is asked to adjust the level of a test tone until it has a specified loudness, either in absolute terms or relative to a standard (e.g., twice as loud, half as loud, etc.) Stevens derived a closed form expression that relates loudness L to intensity I through a constant k as

$$L = kI^{0.3} \qquad (3.37)$$

which states that a doubling of loudness is achieved by a 10 dB difference in intensity level. Stevens defined a "sone", a unit for loudness, as the loudness of a 1 kHz tone at 40 dB SPL. A 1 kHz tone at 50 dB SPL then would have a loudness of 2 sones. Figure 3.15 shows the relation between sones and phons for a 1 kHz tone.

3.7.4 Time Integration

The detection of tones (such as those represented by the absolute threshold of hearing or the equal loudness contours) is also based on the duration of the stimulus tone.

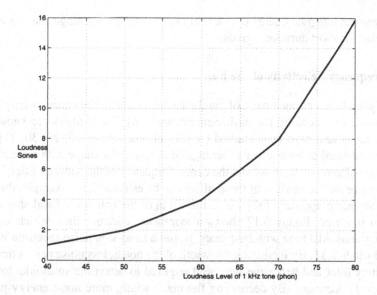

Fig. 3.15. Loudness in sones versus loudness level in phons.

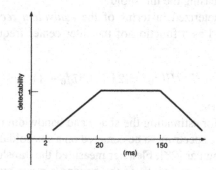

Fig. 3.16. The detectability of a 1 kHz tone as a function of the tone duration in milliseconds.

The relation between duration of the tone, t, and threshold intensity, I, required for detection can be expressed as [26],

$$(I - I_L) \times t = k \tag{3.38}$$

where k is a constant, and I_L is a threshold intensity of a long duration tone pulse.

For example, as shown in [27], the detectability of a tone pulse was constant between 15 to 150 ms, but fell off as the duration increased or decreased beyond these limits as a function of time as shown in Fig. 3.16 for a 1 kHz tone.

The fall in detectability with longer duration indicates that there is a limit to the time over which the ear can integrate energy of the stimulus signal, whereas the fall in detectability at low durations may be connected with the spread of energy over frequency which occurs for signals of short duration. Specifically, it is hypothesized

that the ear can integrate energy over a fairly narrow frequency range and this range
is exceeded for short duration signals.

3.7.5 Frequency Selectivity of the Ear

The peripheral ear acts as a bank of band-pass filters due to the space-to-frequency
transformation induced by the basilar membrane [26]. These filters are known as
auditory filters and have been studied by several researchers [28, 29, 30, 31] and
are conceptualized to have either a rectangular or triangular shape with a simplified
assumption of symmetricity around the center frequency of the auditory filter.

The shape and bandwidth of these filters can be estimated, for example, through
the notched noise approach [32] where the width of the notch of a band-stop noise
spectrum is varied. Figure 3.17 shows a symmetric auditory filter which is centered on a sinusoidal tone with frequency f_0 and a band-stop noise spectrum with a
notch of width $2\Delta f$. By increasing the width of the notch, less noise passes through
the auditory filter and hence the threshold required to detect the sinusoidal tone of
frequency f_0 decreases. By decreasing the notch width, more noise energy passes
through the auditory filter thereby making it harder for the sinusoidal tone to be de-
tected and thereby increasing the threshold.

The filter is parameterized in terms of the *equivalent rectangular bandwidth*
(ERB) and is expressed as a function of the filter center frequency f_0 (expressed
in kHz) as

$$ERB(f_0) = 24.7(4.37f_0 + 1) \tag{3.39}$$

and is shown in Fig. 3.18.

Another approach for estimating the shape and bandwidth of the auditory filters
is by assuming the noise spectrum to be centered on a sinusoidal tone where the filter
is assumed to be rectangular [28]. Fletcher measured the threshold of the sinusoidal
tone as a function of the bandwidth of the bandpass noise by keeping the overall
noise power density constant. It is generally observed that the threshold of the signal

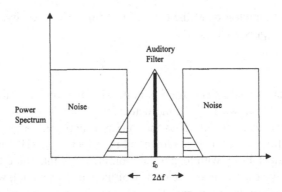

Fig. 3.17. Estimation of the auditory filter shape or bandwidth with the notched noise approach.

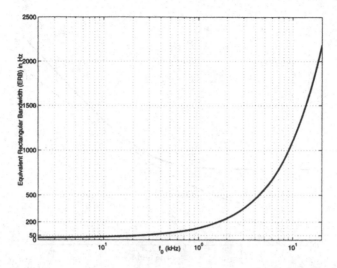

Fig. 3.18. The equivalent rectangular bandwidth in Hz obtained from (3.39).

increases at first as the bandwidth increases, but then flattens out beyond a critical frequency such that any additional increase in noise will not affect detectability of the sinusoidal tone. Fletcher referred to the bandwidth CB where the signal threshold ceased to increase as the *critical bandwidth* or *Bark* scale. The critical bandwidth can be related through one model which is expressed by

$$CB(f) = 25 + 75(1 + 1.4f^2)^{0.69} \tag{3.40}$$

and is shown in Fig. 3.19, where f is in kHz.

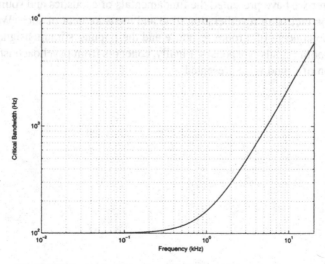

Fig. 3.19. The critical band model obtained from Eq. (3.40).

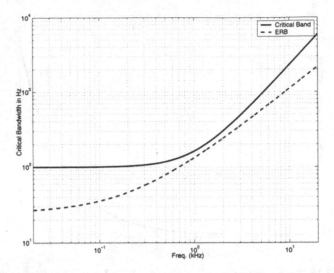

Fig. 3.20. Comparison between ERB and critical bandwidth models.

Figure 3.20 shows the differences between the ERB and critical band filter bandwidths as a function of center frequency. It is evident that the ERB-based auditory filter models have better frequency resolution at lower frequencies than the critical band-based auditory filter models, whereas the differences are generally not substantial at higher frequencies.

3.8 Summary

In this chapter we have presented the fundamentals of acoustics and sound propagation in rooms including reverberation time and measurement of such. We have also presented the concept of a room response and the popular stimulus signals used for measuring room impulse responses. Finally, concepts from psychoacoustics relating to perception of sound were presented.

Part III

Immersive Audio Processing

4

Immersive Audio Synthesis and Rendering Over Loudspeakers

4.1 Introduction

Multichannel sound systems such as those used in movie or music reproduction in 5.1 channel surround sound systems or new formats such as 10.2 channel immersive audio require many more tracks for content production than the number of audio channels used in reproduction. This has been true since the early days of monophonic and two-channel stereo recordings that used multiple microphone signals to create the final one- or two-channel mixes.

In music recording there are several constraints that dictate the use of multiple microphones. These include the sound pressure level of various instruments, the effects of room acoustics and reverberation, the spectral content of the sound source, the spatial distribution of the sound sources in the space, and the desired perspective that will be rendered over the loudspeaker system.

As a result it is not uncommon to find that tens of microphones may be used to capture a realistic musical performance that will be rendered in surround sound. Some of these are placed close to instruments or performers and others farther away so as to capture the interaction of the sound source with the environment.

Despite the emergence of new consumer formats that support multiple audio channels for music, the growth of content has been slow. In this chapter we describe methods that can be used to automatically generate the multiple microphone signals needed for a multichannel rendering without having to record using multiple real microphones, which we refer to as *immersive audio synthesis*. The applications of such virtual microphones can be found in both the conversion of older recordings to today's 5.1 channel formats, but also to upconvert today's 5.1 channel content to future multichannel formats that will inevitably consist of more channels for more realistic reproduction.

Immersive audio rendering involves accurate reproduction of three-dimensional sound fields that preserve the desired spatial location, frequency response, and dynamic range of multiple sound sources in the environment. An immersive audio system is capable of rendering sound images positioned at arbitrary locations around a listener. There are two general approaches to building these systems. The first is to completely surround the listener with a large number of loudspeakers to reproduce the sound field of the target scene. The second is to reproduce the necessary acoustic signals at the ears of the listener as they would occur under natural listening conditions. This method, called binaural audio, is applicable to both headphone and, with some modifications, to loudspeaker reproduction.

In this chapter we describe a general methodology for rendering binaural sound over loudspeakers that can be generalized to multiple listeners. We present the mathematical formulation for the necessary processing of signals to be played from the loudspeakers for the case of two listeners. The methods presented here can also be scaled to more listeners. In theory, loudspeaker rendering of binaural signals requires fewer loudspeakers than the multiple loudspeaker methods proposed by surround sound systems. This is because binaural rendering relies on the reproduction of the sound pressures at each listener's ear. However, binaural rendering over loudspeakers has traditionally been thought of as a single listener solution and has not been applied to loudspeaker systems intended for multiple listeners. There are two general methods for single listener binaural audio rendering that can be categorized as headphone reproduction and loudspeaker reproduction [33, 34, 35, 36].

Head-related binaural recording, or dummy-head stereophony methods, attempt to accurately capture and reproduce at each eardrum the sound pressure generated by sound sources and their interactions with the acoustic environment and the pinnae, head, and torso of the listeners [37, 38, 39]. Transaural audio is a method that was developed to deliver binaural signals to the ears of listeners using two loudspeakers. The basic idea is to filter the binaural signal such that the crosstalk terms from the loudspeakers to the opposite side ear are reduced so that the signals at the ears of the listener approach those captured by a binaural recording. The technique was presented in [41, 42] and later developed fully by Cooper and Bauck [43], who coined the term "transaural audio". Previous work in the literature [43, 44, 45, 46] has focused on both theoretical and practical methods for generalizing crosstalk cancellation filter design using matrix formulations. Cooper and Bauck [46] also discussed some ideas for developing transaural systems for multiple listeners with multiple loudspeakers.

Crosstalk cancellation filter design for loudspeaker reproduction systems has been proposed using least mean squares (LMS) adaptive algorithm methods in symmetric or nonsymmetric environments [44, 45, 51, 52]. As the eigenvalue spread of the input autocorrelation matrix increases, the convergence speed of the LMS adaptive algorithm for multichannel adaptation decreases. To solve this problem, algorithms such as discrete Fourier transform (DFT)/LMS and discrete cosine transform (DCT)/LMS can be used to decorrelate the input signals by preprocessing with a transformation that is independent of the input signal. In this chapter we present crosstalk cancellation filter design methods based on the LMS adaptive inverse algorithm, and the normalized frequency domain adaptive filter (NFDAF) LMS inverse algorithm [55, 54]. The authors wish to acknowledge Dr. Athanasios Mouchtaris whose PhD dissertation at the USC Immersive Audio Laboratory formed the basis for much of the work described regarding synthesis and Dr. Jong-Soong Lim whose PhD dissertation formed the basis for much of the work on rendering.

4.2 Immersive Audio Synthesis

4.2.1 Microphone Signal Synthesis

The problem of synthesizing a virtual microphone signal from a real microphone signal recorded at a different position in the room can be formulated as a general filtering problem. In order to derive the filter, it is first necessary to train the system using a pair of real microphones at the desired locations. If we call the signals in these microphones m_1 and m_2 then it is desirable to construct a filter V that when applied to m_1 results in a signal that is as close as possible to m_2. The difference between the synthesized signal and the real signal must be as small as possible, both from an objective measure as well as from a psychoacoustic point of view.

There are several possible methods that can be used to find the necessary filters that will synthesize the virtual microphone signals. The most common among these methods may be to use an adaptive filter approach. The drawback of such a choice in this case is that acoustical performance spaces used for recording may exhibit very long reverberation times (sometimes longer than two seconds) and this would impose unreasonably large tap and associated computational and memory requirements for finite impulse response filters derived from adaptive methods.

In the sections below, we describe algorithms that utilize infinite impulse response filters. This is a particularly relevant choice for synthesizing virtual microphones that are used to capture the interactions of sound sources with the acoustical characteristics of the space and thus are placed at large distances from the sound sources. In a way, these microphones represent the reverberant field that has been modeled extensively using comb-filter techniques based on IIR. The resulting virtual microphone synthesis filters are more computationally efficient than their FIR counterparts.

To ensure that the resulting filters are stable, it is important to follow a design approach that results in a minimum-phase IIR filter. Using a classical music recording in a concert hall as an example, we can define the direct signal from the orchestra on stage as s and m_1 and m_2 as the recorded microphone signals that are captured in two microphones. In effect, these signals are the result of convolution of the dry orchestra signal with the room impulse response for each microphone position. The method described below is based on an all-pole model of these filters. That is, the filtering between s and m_1 can be modeled as an all-pole filter, resulting in filter A_1 and the filtering between s and m_2 as another all-pole model resulting in filter A_2. Then, the desired filter V can be applied to the signal m_1 to generate m_{2p}, which is an estimate of m_2. The filter $V = A_1/A_2$, is a stable IIR filter.

We can consider each of the microphone signals to be wide-sense stationary, or at least, that they can be separated into blocks that are wide-sense stationary. Under that assumption, we can model each signal as an autoregressive process. If the direct signal from the source is denoted by S in the frequency domain ($s(n)$ in the time domain), and the room impulse responses from the source to each microphone are denoted as V_1 and V_2, respectively, then $M_1 = V_1 S$ and $M_2 = V_2 S$.

Then, V_1 and V_2 can be modeled as all-pole filters of order r resulting in the following AR models

$$s_1(n) = \sum_{i-1}^{r} a_1(i)m_1(n-i) + s(n)$$

$$s_2(n) = \sum_{i-1}^{r} a_2(i)m_2(n-i) + s(n) \tag{4.1}$$

These can be expressed in the z-domain as

$$V_1(n) = \frac{S_1(z)}{S(z)} = \frac{1}{A_1(z)}$$

$$V_2(n) = \frac{S_2(z)}{S(z)} = \frac{1}{A_2(z)} \tag{4.2}$$

in which the denominator terms in the all-pole filters are given by

$$A_1(z) = 1 - \sum_{k=1}^{r} a_1(n)z^{-k}$$

$$A_2(z) = 1 - \sum_{k=1}^{r} a_2(n)z^{-k} \tag{4.3}$$

The required filter V, which can be used to synthesize the microphone signal m_2 from the reference signal m_1, is

$$V = \frac{V_2}{V_1} = \frac{A_1}{A_2} \tag{4.4}$$

One way to ensure that each virtual microphone filter V is both stable and computationally efficient, is to use linear prediction analysis to design a stable all-pole filter. With linear prediction, a certain number of past samples in each signal time domain record are linearly combined to provide an estimate of future samples. For example, if at time n in the signal m_1, q past samples are considered then the estimate of the signal at time n can be written as

$$m_1^{lp}(n) = \sum_{k=1}^{q} a(i)m_1(n-k) \tag{4.5}$$

The prediction error of this process from the actual microphone signal is then

$$e(n) = m_1(n) - m_1^{lp}(n) \tag{4.6}$$

which can be written as a transfer function in the z-domain

$$\frac{E(z)}{M_1(z)} = 1 - \sum_{k=1}^{q} a(k)z^{-k} \tag{4.7}$$

In real-world situations, an autoregressive model will not provide an exact model of the system and will result in a modeling error term given by

$$e(n) = s(n) + e_{AR}(n) \tag{4.8}$$

in which $e_{AR}(n)$ is the error that arises from the incorrect modeling of the source. Minimizing the error $e(n)$ is equivalent to minimizing the error $e_{AR}(n)$. Furthermore, minimization of the error $e(n)$ produces the coefficients $a(i)$ in (4.7) that in fact are the same as those of filter A_1. This is easily seen if $E(z)$ is substituted by $S(z)$ in (4.7). These coefficients can be calculated using linear prediction to minimize the mean squared error between $m_1(n)$ and $m_1^{lp}(n)$ [10, 8]. Linear prediction is a special case of linear optimum filtering, thus the principle of orthogonality holds. Accordingly, minimization of the error is equivalent to the error $e(n)$ being orthogonal to all the input samples $m_1(n - k)$ from which the error at time n is calculated (such that k lies in the interval $[1, q]$); that is,

$$E(m_1(n - k)e(n)) = 0 \tag{4.9}$$

$$E(m_1(n - k)(m_1(n) - \sum_{k=1}^{q} a(i)m_1(n - i))) = 0 \tag{4.10}$$

$$r(-k) = \sum_{k=1}^{q} a(i)r(i - k) \tag{4.11}$$

in which $r(n)$ is the autocorrelation function of $m_1(n)$. Equation (4.11) makes use of the fact that the process m_1 is wide-sense stationary in the block under consideration. Finally, because the autocorrelation function is symmetric, and the absolute value of $i - k$ is in the interval $[0, q - 1]$ (4.11) can be rewritten in matrix form as

$$\begin{bmatrix} r(0) & r(1) & \dots & r(q-1) \\ r(1) & r(0) & \dots & r(q-2) \\ \vdots & \vdots & \ddots & \vdots \\ r(q-1) & r(q-2) & \dots & r(0) \end{bmatrix} \begin{bmatrix} a(1) \\ a(2) \\ \vdots \\ a(q) \end{bmatrix} = \begin{bmatrix} r(1) \\ r(2) \\ \vdots \\ r(q) \end{bmatrix} \tag{4.12}$$

The coefficients $a(i)$ of the virtual microphone filter can be found from the above equation by inverting the correlation matrix $\mathbf{R}$. This can be performed very efficiently using a recursive method such as the Levinson and Durbin algorithm because of the form of the correlation matrix $\mathbf{R}$ and under the assumption of ergodicity.

4.2.2 Subjective Evaluation of Virtual Microphone Signals

The methods described in the previous section must be applied in blocks of data of the two microphone signal processes m_1 and m_2. A set of experiments was conducted to subjectively verify the validity of these methods. Signal block lengths of 100,000 samples were chosen because the reverberation time of the hall from which the recordings were obtained is 2 sec. and were sampled at 48 kHz. Experiments were performed with various orders of filters A_1 and A_2 to obtain an understanding of the tradeoffs between performance and computational efficiency. Relatively high orders were required to synthesize a signal m_2 from m_1 with an acceptable error between m_{2p} (the reproduced process) and m_2 (the actual microphone signal). The error was assessed through blind A/B/X listening evaluations. An order of 10,000 coefficients for both the numerator and denominator of V resulted in an error between the original and synthesized signals that was not detectable by the listeners. The performance of the filter was also evaluated by synthesizing blocks from a section of the signal different from the one that was used for designing the filter. Again, the A/B/X evaluation showed that for orders higher than 10,000 the synthesized signal was indistinguishable from the original. Although such high order filters are impractical for real-time applications, the performance of this method is an indication that the model is valid and therefore worthy of further investigation to achieve filter optimization.

In addition to listening evaluations, a mathematical measure of the distance between the synthesized and the original processes can be found. This measure can be used during the optimization process in order to achieve good performance and at the same time minimize the number of coefficients. The difficulty in defining such a measure is that it must also be psychoacoustically valid. This problem has been addressed in speech processing in which measures such as the log spectral distance and the Itakura distance are used [47]. In the case presented here, the spectral characteristics of long sequences must be compared with spectra that contain a large number of peaks and dips that are narrow enough to be imperceptible to the human ear. To approximately match the spectral resolution of the human ear 1/3 octave smoothing was performed [26] followed by a comparison of the resulting smoothed spectral cues. The results are shown in Fig. 4.1 where the error match between the spectra of the original (measured) microphone signal and the synthesized signal are compared. The two spectra are practically indistinguishable below 10 kHz. Although the error increases somewhat at higher frequencies, the listening evaluations show that this is not perceptually significant.

4.2.3 Spot Microphone Synthesis Methods

The method described in the previous section is appropriate for the synthesis of microphones placed far from the source that capture mostly reverberant sound in the recording environment. However, it is common practice in music recording to also use microphones that are placed very close to individual instruments. Synthesis of such virtual microphone signals requires a different approach because these

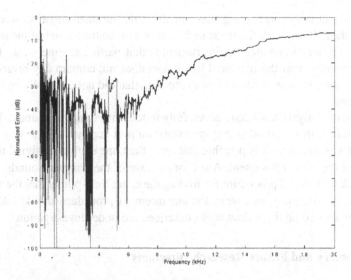

Fig. 4.1. Magnitude response error between the approximating and approximated spectra.

signals exhibit quite different spectral characteristics compared to the reference microphones. These microphones are used, for example, near the tympani or the woodwinds in classical music so that these instruments can be emphasized in the multichannel mix during certain passages. The signal in these microphones is typically not reverberant because of their proximity to the instruments.

As suggested in the previous section, this problem can be classified as a system identification problem. The most important consideration in this case is that it is not theoretically possible to design a generic time-invariant filter that will be suitable for any recording. Such a filter would have to vary with the temporal characteristics of the frequency response of the signal. The response is closely related to the joint time and frequency properties of the reference microphone signals.

The approach that we followed for recreating these virtual microphones is based on a method used for synthesizing percussive instrument sounds [48]. Thus, the method described here is applicable only for microphones located near percussion instruments. According to [48], it is possible to synthesize percussive sounds in a natural way, by following an excitation/filter model. The excitation part corresponds to the interaction between the exciter and the resonating body of the instrument and lasts until the structure reaches a steady vibration, and the resonance part corresponds to the free vibration of the instrument body. The resonance part can be easily described from the frequency response of the instrument using several modeling methods (e.g., the AR modeling method that was described in the previous paragraph). Then, the excitation part can be derived by filtering the instrument's response with the inverse of the resonance filter. The excitation part is independent of the frequencies and decays of the harmonics of the instrument at a given time (after the instrument has reached a steady vibration) so it can be used for synthesizing different sounds by using an appropriate resonance filter. Therefore, it is possible to derive an excitation signal from a recording that contains only the instrument we wish to enhance and then filter it with the resonance filter at a given time point of the reference recording in order to enhance the instrument at that particular time point. It is important to mention that the recreated instrument does not contain any reverberation if the excitation part was derived from a recording that did not originally contain any reverberation.

The above analysis has been successfully tested with tympani sounds. A potential drawback of this method is that the excitation part depends on the way that the instrument was struck so it is possible that more than one excitation signal might be required for the same instrument. Also, for the case of the tympani sounds, it is not an easy task to define a procedure for finding the exact time points that the tympani was struck, that is, the points when the enhancement procedure should take place. Solutions to overcome these drawbacks described are under investigation.

4.2.4 Summary and Future Research Directions

The methods described above are effective for synthesizing signals in virtual microphones that are placed at a distance from the sound source (e.g., orchestra) and therefore, contain more reverberation. The IIR filtering solution was proposed exactly for addressing the long reverberation-time problem, which meant long impulse responses for the filters to be designed. On the other hand, signals from microphones located close to individual sources (e.g., spot microphones near a particular musical instrument) do not contain very much reverberation. A completely different problem arises when trying to synthesize such signals. Placing such microphones near individual sources with varying spectral characteristics results in signals whose frequency content will depend highly on the microphone positions.

In order to synthesize signals in such closely placed microphones it is necessary to identify the frequency bands that need to be amplified or attenuated for each microphone. This can be easily achieved when the reference microphone is relatively far from the orchestra, so that we can consider that all frequency bands were equally weighted during the recording. In order to generate a reference signal from such a distant microphone that can be used to synthesize signals in the nonreverberant microphones it is necessary to find some method for dereverberating the reference signal.

One complication with this approach is that we do not know the filter that transforms the signal from the orchestra to the reference microphone. We are investigating methods for estimating these filters based on a technique for blind channel identification using cross-spectrum analysis [49]. The idea is to use the two closely spaced microphones in the center (hanging above the conductor's head) as two different observations of the same signal processed by two different channels (the path from the orchestra to each of the two microphones). The Pozidis and Petropulu algorithm uses the phase of the cross-spectrum of the two observations and allows us to estimate the two channels. Further assumptions, though, need to be made in order to have a unique solution to the problem. The most important is that the two channels are assumed to be of finite length. In general, however, they can be nonminimum phase, which is a desired property. All the required assumptions are discussed in [49] and their implications for the specific problem examined here are currently under investigation. After the channels have been identified, the recordings can be equalized using the estimated filters. These filters can be nonminimum phase, as explained earlier, so a method for equalizing nonminimum phase channels must be used. Several methods exist for this problem; see, for example, [50]. The result will be not only a dereverberated signal but an equalized signal, ideally equal to the signal that the microphone would record in an anechoic environment. That signal could then be used as the seed to generate virtual microphone signals that would result in multichannel mixes simulating various recording venues.

4.3 Immersive Audio Rendering

4.3.1 Rendering Filters for a Single Listener

A typical two-loudspeaker listening situation is shown in Fig. 4.2, in which X_L and X_R are the binaural signals sent to listener's ears E_L and E_R through loudspeakers S_1 and S_2.

The system can be fully described by the following matrix equation

$$\begin{bmatrix} E_L \\ E_R \end{bmatrix} = \begin{bmatrix} T_1 & T_2 \\ T_3 & T_4 \end{bmatrix} \cdot \begin{bmatrix} S_1 \\ S_2 \end{bmatrix} = \begin{bmatrix} T_1 & T_2 \\ T_3 & T_4 \end{bmatrix} \cdot \mathbf{W} \cdot \begin{bmatrix} X_L \\ X_R \end{bmatrix} \qquad (4.13)$$

in which W is the matrix of the crosstalk canceller, and T_1, T_2, T_3, and T_4 are the head-related transfer functions (HRTFs) between the loudspeakers and ears. To generate a spatially rendered sound image, a rendering filter is required that delivers the left channel binaural signal X_L to E_L, and the right channel binaural signal X_R to E_R, while simultaneously eliminating unwanted crosstalk terms. If the above conditions are satisfied exactly, matrix equation (4.13) can be formulated as follows

$$\mathbf{E} = \mathbf{T} \cdot \mathbf{W} \cdot \mathbf{X} = \mathbf{X} \qquad (4.14)$$

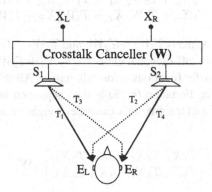

Fig. 4.2. Geometry and signal paths from input binaural signals to ears that show the ipsilateral signal paths (T_1, T_4) and contralateral signal paths (T_2, T_3).

In (4.14), the listener's ear signals and input binaural signals are $\mathbf{E} = [E_L E_R]^T$, $\mathbf{X} = [X_L X_R]^T$, respectively. The rendering system transfer function matrix $\mathbf{T}$ is

$$\mathbf{T} = \begin{bmatrix} T_1 & T_2 \\ T_3 & T_4 \end{bmatrix} \tag{4.15}$$

To obtain optimum performance and deliver the desired signal to each ear, the matrix product of transfer function matrix $\mathbf{T}$ and crosstalk canceling weight vector matrix $\mathbf{W}$ should be the identity matrix

$$\mathbf{T} \cdot \mathbf{W} = \begin{bmatrix} 1 & 0 \\ 0 & 1 \end{bmatrix} \tag{4.16}$$

Therefore the generalized rendering filter weight matrix $\mathbf{W}$ requires four weight vectors to produce the desired signal at the ears of a single listener.

Time Domain Adaptive Inverse Control Filter

The weight vector matrix $\mathbf{W}$ described above can be implemented using the least mean squares adaptive inverse algorithm [51]. Matrix equations (4.14) and (4.16) must be modified based on the adaptive inverse algorithm for multiple channels as follows [44, 45, 52]

$$\mathbf{E} = \mathbf{T} \cdot \mathbf{W} \cdot \mathbf{X} = \mathbf{T} \cdot \begin{bmatrix} W_1 & W_3 \\ W_2 & W_4 \end{bmatrix} \cdot \mathbf{X} = \mathbf{X} \tag{4.17}$$

The desired result is to find $\mathbf{W}$ so that it cancels the crosstalk signals perfectly. Then the signals $\mathbf{E}$ arriving at the ears are exactly the same as the input binaural signals $\mathbf{X}$. Equation (4.17) can be written as

$$\begin{aligned} \mathbf{E} &= \begin{bmatrix} T_1 W_1 + T_2 W_2 & T_1 W_3 + T_2 W_4 \\ T_3 W_1 + T_4 W_2 & T_3 W_3 + T_4 W_4 \end{bmatrix} \cdot \mathbf{X} \\ &= \begin{bmatrix} T_1 W_1 X_L + T_2 W_2 X_L + T_1 W_3 X_R + T_2 W_4 X_R \\ T_3 W_1 X_L + T_4 W_2 X_L + T_3 W_3 X_R + T_4 W_4 X_R \end{bmatrix} \end{aligned} \tag{4.18}$$

in which, the diagonal elements $T_1 W_1 + T_2 W_2$ and $T_3 W_3 + T_4 W_4$ are the ipsilateral transfer functions, and the off-diagonal elements $T_1 W_3 + T_2 W_4$ and $T_3 W_1 + T_4 W_2$ are the contralateral transfer functions (crosstalk terms). All the vectors in (4.18) are in the frequency domain. Equation (4.18) is then separated into two matrices: one that is the matrix product of the crosstalk canceller weight vectors and the other with the remaining terms

$$\mathbf{E} = \begin{bmatrix} T_1 X_L & T_2 X_L & T_1 X_R & T_2 X_R \\ T_3 X_L & T_4 X_L & T_3 X_R & T_4 X_R \end{bmatrix} \cdot \mathbf{W}' = \mathbf{X} \tag{4.19}$$

in which $\mathbf{W}'$ is a column matrix $[W_1 W_2 W_3 W_4]^T$. The block diagram is shown in Fig. 4.3.

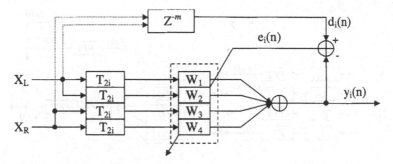

Fig. 4.3. LMS block diagram for the estimation of the crosstalk cancellation filter with $d_i(n) = X_L(n - m)$, and $d_i(n) = X_R(n - m)$ for the left and right channels, respectively.

Using the time domain LMS adaptive algorithm, the weight vectors are updated as follows,

$$W_i(n + 1) = W_i(n) + \mu(-\hat{\nabla}_i(n)), \qquad i = 1, \ldots, 4 \qquad (4.20)$$

The positive scalar step size μ controls the convergence rate and steady-state performance of the algorithm. The gradient estimate, $\hat{\nabla}(n)$, is simply the derivative of $e^2(n)$ with respect to $W(n)$ [56]. Therefore gradient estimates in the time domain can be found as

$$\hat{\nabla}_i(n) = -2\{e_1(n)[T_i(n) * X_L(n)] + e_2(n)[T_{i+2}(n) * X_L(n)]\} \qquad i = 1, 2$$
$$\hat{\nabla}_i(n) = -2\{e_1(n)[T_{i-2}(n) * X_R(n)] + e_2(n)[T_i(n) * X_R(n)]\} \qquad i = 3, 4$$
$$(4.21)$$

in which all input binaural signals and transfer functions are time domain sequences. The output error is given by

$$e_i(n) = d_i(n) - y_i(n)$$
$$= X_i(n - m) - \{[W_1(n) * T_{2i-1}(n) + W_2(n) * T_{2i}(n)] * X_L(n)$$
$$+ [W_3(n) * T_{2i-1}(n) + W_4(n) * T_{2i}(n)] * X_R(n)\} \qquad i = 1, 2 \quad (4.22)$$

in which $X_i(n)$ is

$$X_i(n) = \begin{cases} X_L(n) & i = 1 \\ X_R(n) & i = 2 \end{cases} \qquad (4.23)$$

Figure 4.3 shows that $d_i(n)$ could simply be a pure delay, say of m samples, which will assist in the equalization of the minimum phase components of the transfer function matrix in (4.18). The inclusion of an appropriate modeling delay significantly reduces the mean square error produced by the equalization process. The filter length, as well as the delay m, can be selected based on the minimization of the mean squared error. This method can be used either offline or in real-time according to the location of the virtual sound source and the position of the listener's head. The weight vectors of the crosstalk canceller can be chosen to be either an FIR or an IIR filter.

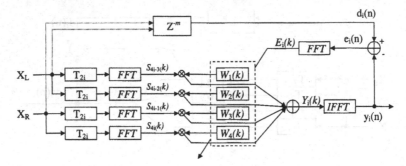

Fig. 4.4. Frequency domain adaptive LMS inverse algorithm block diagrams using overlap-save method for the estimation of crosstalk canceller weighting vectors based on Fig. 4.3 ($i = 1, 2$).

Frequency Domain Adaptive Inverse Filter

Frequency domain implementations of the LMS adaptive inverse filter have several advantages over time domain implementations that include improved convergence speed and reduced computational complexity. In practical implementations of frequency domain LMS adaptive filters, the input power varies dramatically over the different frequency bins. To overcome this, the frequency domain adaptive filter (FDAF) LMS inverse algorithm [59] can be used to estimate the input power in each frequency bin. The power estimate can be included directly in the frequency domain LMS algorithm [55]. The adaptive inverse filter algorithm shown in Fig. 4.3 is modified in the frequency domain using the overlap-save method FDAF LMS inverse algorithm [54], which is shown in Fig. 4.4.

The general form of FDAF LMS algorithms can be expressed as follows,

$$W(k+1) = 2\mu(k)X^H(k)E(k) \tag{4.24}$$

in which the superscript H denotes the complex conjugate transpose. The time-varying matrix $\mu(k)$ is diagonal and it contains the step sizes $\mu_1(k)$. Generally, each step size is varied according to the signal power in that frequency bin l. In the crosstalk canceller implementation described here

$$W_i(k+1) = W_i(k) + \mu \times F\left\{F^{-1}\left[\frac{S_i^H(k)}{P_i(k)} \cdot E_1(k) + \frac{S_{4+i}^H(k)}{P_{4+i}(k)} \cdot E_2(k)\right]\right\}$$
$$i = 1, \dots, 4 \tag{4.25}$$

In (4.25), μ is a fixed scalar and $S_i(k)$ is a product of input signal X_L or X_R and the transfer function shown in Fig. 4.4. $P_i(k)$ is an estimation of the signal power in the nth-input signal

$$P_i(k) = \lambda P_i(k-1) + \alpha|S_i(k)|^2 \tag{4.26}$$

in which $\lambda = 1 - \alpha$ is a forgetting factor. $P_i(k)$ and $S_i(k)$ are vectors composed of N different bins.

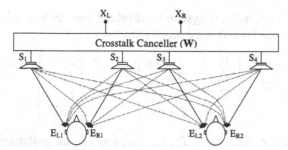

Fig. 4.5. Geometry for four loudspeakers and two listeners with the ipsilateral signal paths (solid lines) and contralateral undesired signal paths (dotted lines).

4.3.2 Rendering Filters for Multiple Listeners

Filters for rendering immersive audio to multiple listeners simultaneously can be implemented using the filter design methods described above. For the case of two listeners with four loudspeakers, the crosstalk canceller weighting vectors for the necessary FIR filters can be determined using the least mean squares adaptive inverse algorithm [8, 58] in which the adaptation occurs in the sampled time domain and in the frequency domain. The purpose of virtual loudspeaker rendering for two listeners is to generate virtual sound sources at two listener's ears. In order to deliver the appropriate sound field to each ear, it is necessary to eliminate crosstalk signals that are inherent in all loudspeaker-based systems (Fig. 4.5).

For the case of two listeners and four loudspeakers there exist 12 crosstalk paths that should be removed. A method is presented here for implementing such a system based on adaptive algorithms. Results from four different configurations of adaptive filter implementations are discussed that can deliver a binaural audio signal to each listener's ears for four different combinations of geometry and direction of rendered image. The basic configuration is shown in Fig. 4.5.

General NonSymmetric Case

The position of each listener's head is assumed to be at an arbitrary position relative to each loudspeaker pair $S_1 - S_2$ and $S_3 - S_4$. Therefore 16 possible head-related transfer functions can be generated from each of four loudspeakers to two listeners (denoted as $T_1 \sim T_{16}$ as shown in Fig. 4.6).

The purpose of this rendering filter is to deliver left-channel and right-channel audio signals X_L and X_R to each listener's ears, respectively, with both listeners perceiving the same spatial location for the rendered image. Transfer functions in Fig. 4.6 are formulated into matrix equations as

$$\mathbf{E} = \mathbf{T} \cdot \mathbf{S} = \mathbf{T} \cdot \mathbf{W} \cdot \mathbf{X} \qquad (4.27)$$

in which the ear signal matrix $\mathbf{E}$ is $\mathbf{E} = [E_{L1} E_{R1} E_{L2} E_{R2}]^T$, the loudspeaker signal matrix $\mathbf{S}$ is $\mathbf{S} = [S_1 S_2 S_3 S_4]^T$, and the input binaural signal matrix $\mathbf{X}$ is

$\mathbf{X} = [X_L X_R]^T$. The rendering system head-related transfer function matrix $\mathbf{T}$ and crosstalk cancellation filter $\mathbf{W}$ are defined as follows

$$\mathbf{T} = \begin{bmatrix} T_1 & T_2 & T_3 & T_4 \\ T_5 & T_6 & T_7 & T_8 \\ T_9 & T_{10} & T_{11} & T_{12} \\ T_{13} & T_{14} & T_{15} & T_{16} \end{bmatrix} \quad \mathbf{W} = \begin{bmatrix} T_1 & T_5 \\ T_2 & T_6 \\ T_3 & T_7 \\ T_4 & T_8 \end{bmatrix} \tag{4.28}$$

From (4.28), the signal paths T_1, T_6, T_{11}, and T_{16} are ipsilateral signal paths, and $T_2, T_3, \ldots, T_{15}$ are undesired contralateral crosstalk signal paths. If the product of matrices $\mathbf{T}$, $\mathbf{W}$, and $\mathbf{X}$ can be simplified as (4.29) for the same side rendered image, the crosstalk cancellation filter $\mathbf{W}$ will be the optimum inverse control filter. Therefore the desired binaural input signals X_L and X_R can be delivered to the ears of each listener without crosstalk.

$$\mathbf{E} = \mathbf{T} \cdot \mathbf{W} \cdot \mathbf{X} = \begin{bmatrix} 1 & 0 \\ 0 & 1 \\ 1 & 0 \\ 0 & 1 \end{bmatrix} \mathbf{X} = \begin{bmatrix} X_L \\ X_R \\ X_L \\ X_R \end{bmatrix} \tag{4.29}$$

Equation (4.27) is modified as follows

$$\mathbf{E} = \mathbf{GX} \tag{4.30}$$

$$\mathbf{G} = \begin{bmatrix} T_1W_1 + T_2W_2 + T_3W_3 + T_4W_4 & T_1W_5 + T_2W_6 + T_3W_7 + T_4W_8 \\ T_5W_1 + T_6W_2 + T_7W_3 + T_8W_4 & T_5W_5 + T_6W_6 + T_7W_7 + T_8W_8 \\ T_9W_1 + T_{10}W_2 + T_{11}W_3 + T_{12}W_4 & T_9W_5 + T_{10}W_6 + T_{11}W_7 + T_{12}W_8 \\ T_{13}W_1 + T_{14}W_2 + T_{15}W_3 + T_{16}W_4 & T_{13}W_5 + T_{14}W_6 + T_{15}W_7 + T_{16}W_8 \end{bmatrix} .$$

Equation (4.30) can then be modified as

$$\mathbf{E} = \begin{bmatrix} T_1X_L & T_2X_L & T_3X_L & T_4X_L & T_1X_R & T_2X_R & T_3X_R & T_4X_R \\ T_5X_L & T_6X_L & T_7X_L & T_8X_L & T_5X_R & T_6X_R & T_7X_R & T_8X_R \\ T_9X_L & T_{10}X_L & T_{11}X_L & T_{12}X_L & T_9X_R & T_{10}X_R & T_{11}X_R & T_{12}X_R \\ T_{13}X_L & T_{14}X_L & T_{15}X_L & T_{16}X_L & T_{13}X_R & T_{14}X_R & T_{15}X_R & T_{16}X_R \end{bmatrix} \cdot \mathbf{W}$$

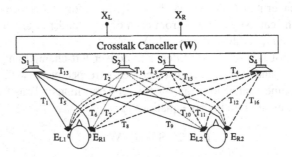

Fig. 4.6. Geometry and transfer functions for four loudspeakers and two listeners (general nonsymmetric case).

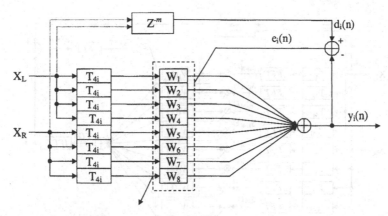

Fig. 4.7. LMS block diagrams for the estimation of crosstalk canceller weighting vectors in the general nonsymmetric case, with $d_i(n) = X_L(n - m)$ and $d_i(n) = X_R(n - m)$ for the left and right channels, respectively.

$$= \begin{bmatrix} X_L & X_R & X_L & X_R \end{bmatrix}^T \qquad (4.31)$$

in which $\mathbf{W}'$ is a column matrix $[W_1 W_2 W_3 W_4 W_5 W_6 W_7 W_8]^T$. Figure 4.7 shows a block diagram of weight vectors generated using the LMS algorithm for the general non-symmetric case.

The weight vectors of the crosstalk canceller are updated based on the LMS adaptive algorithm

$$W_i(n+1) = W_i(n) + \mu \times (-\hat{\nabla}_i(n)) \qquad i = 1, \dots, 8 \qquad (4.32)$$

In (4.30), the convergence rate of the LMS adaptive algorithm is controlled by step size μ. The gradient estimate, $\hat{\nabla}(n)$, is simply the derivative of $e^2(n)$ with respect to $W(n)$. Therefore the gradient estimates in the time domain are

$$\hat{\nabla}_i(n) = -2\{e_1(n)[T_i(n) * X_L(n)] + e_2(n)[T_{4+i}(n) * X_L(n)]$$
$$+ e_3(n)[T_{8+i}(n) * X_L(n)] + e_4(n)[T_{12+i}(n) * X_L(n)]\} \qquad i = 1, \dots, 4$$
$$\hat{\nabla}_i(n) = -2\{e_1(n)[T_{i-4}(n) * X_R(n)] + e_2(n)[T_i(n) * X_R(n)]$$
$$+ e_3(n)[T_{4+i}(n) * X_R(n)] + e_4(n)[T_{8+i}(n) * X_R(n)]\} \qquad i = 5, \dots, 8$$
$$(4.33)$$

In (4.33), all input binaural signals and transfer functions are sample sequences in the time domain. The output error is given by

$$e_i(n) = d_i(n) - y_i(n) = X_i(n - m) - y_i(n) \qquad i = 1, \dots, 4 \qquad (4.34)$$

in which $X_i(n)$ is

$$X_i(n) = \begin{cases} X_L(n) & i = 1, 3 \\ X_R(n) & i = 2, 4 \end{cases} \qquad (4.35)$$

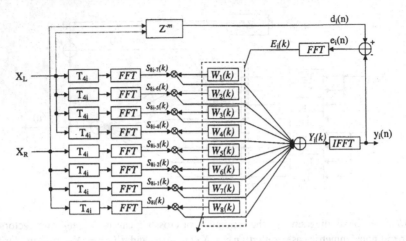

Fig. 4.8. Frequency domain adaptive LMS inverse algorithm block diagrams using overlap-save method for the estimation of crosstalk canceller weighting vectors based on Fig. 4.7 ($i = 1, \ldots, 4$).

The filter output $y(n)$ is

$$
\begin{aligned}
y_i(n) &= [W_1(n) * T_{4i-3}(n) + W_2(n) * T_{4i-2}(n) + W_3(n) * T_{4i-1}(n) \\
&\quad + W_4(n) * T_{4i}(n)] * X_L(n) \\
&= [W_5(n) * T_{4i-3}(n) + W_6(n) * T_{4i-2}(n) + W_7(n) * T_{4i-1}(n) \\
&\quad + W_8(n) * T_{4i}(n)] * X_R(n) \qquad i = 1, \ldots, 4 \qquad (4.36)
\end{aligned}
$$

Figure. 4.8 shows the block diagram of the frequency domain adaptive LMS inverse algorithm for the general nonsymmetric case.

By using the weight vector adaptation algorithm in (4.24),

$$
\begin{aligned}
W_i&(n+1) \\
&= W_i(n) + \mu \times fft \left\{ ifft \left[\frac{S_i^H(k)}{P_i(k)} \cdot E_1(k) + \frac{S_{8+i}^H(k)}{P_{8+i}(k)} \cdot E_2(k) \right. \right. \\
&\qquad \left. \left. + \frac{S_{16+i}^H(k)}{P_{16+i}(k)} \cdot E_3(k) + \frac{S_{24+i}^H(k)}{P_{24+i}(k)} \cdot E_4(k) \right] \right\} \\
&\qquad\qquad\qquad\qquad\qquad\qquad i = 1, \ldots, 8 \qquad (4.37)
\end{aligned}
$$

In (4.37), $S_i(k)$ and $P_i(k)$ are described in section B for single listener case.

Symmetric Case

Each of the two listeners in this configuration is seated at the center line of each loudspeaker pair. This implies that several of the HRTFs in this geometry are identical due to symmetry (assuming that no other factors such as room acoustics influence the system). Therefore the 16 HRTFs from T1 to T16 can be reduced to just 6 HRTFs

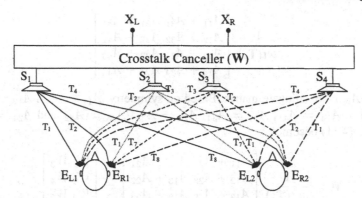

Fig. 4.9. Geometry and transfer functions for four loudspeakers and two listeners with symmetry geometry.

$(T_1 = T_6 = T_{11} = T_{16}, T_2 = T_5 = T_{12} = T_{15}, T_3 = T_{14}, T_4 = T_{13}, T_7 = T_{10}$, and $T_8 = T_9)$ as shown in Fig. 4.9.

This filter has the same property as the general nonsymmetric case. But the eight filters $(W_1 \sim W_8)$ for the general case can be reduced to four filters $(W_1 \sim W_4)$ using the symmetry property. From (4.27), (4.29), and the symmetry property we find

$$\mathbf{T}_s \cdot \mathbf{W} = \begin{bmatrix} 1 & 0 \\ 0 & 1 \\ 1 & 0 \\ 0 & 1 \end{bmatrix} \tag{4.38}$$

The symmetric rendering system HRTFs matrix T_s is

$$\mathbf{T}_s = \begin{bmatrix} T_1 & T_2 & T_3 & T_4 \\ T_2 & T_1 & T_7 & T_8 \\ T_8 & T_7 & T_1 & T_2 \\ T_4 & T_3 & T_2 & T_1 \end{bmatrix} \tag{4.39}$$

Therefore the crosstalk canceller weighting vector matrix $\mathbf{W}$ will be

$$\mathbf{W} = \mathbf{T}_s^{-1} \cdot \begin{bmatrix} 1 & 0 \\ 0 & 1 \\ 1 & 0 \\ 0 & 1 \end{bmatrix} \tag{4.40}$$

From (4.40), the matrix inverse is expanded as follows:

$$\mathbf{W} = \frac{1}{det(\mathbf{T}_s)} \begin{bmatrix} A_{11} & A_{21} & A_{31} & A_{41} \\ A_{12} & A_{22} & A_{32} & A_{42} \\ A_{13} & A_{23} & A_{33} & A_{43} \\ A_{14} & A_{24} & A_{34} & A_{44} \end{bmatrix} \cdot \begin{bmatrix} 1 & 0 \\ 0 & 1 \\ 1 & 0 \\ 0 & 1 \end{bmatrix}$$

$$= \frac{1}{det(\mathbf{T}_s)} \begin{bmatrix} A_{11} + A_{31} & A_{21} + A_{41} \\ A_{12} + A_{32} & A_{22} + A_{42} \\ A_{13} + A_{33} & A_{23} + A_{43} \\ A_{14} + A_{34} & A_{24} + A_{44} \end{bmatrix} \tag{4.41}$$

in which A_{ij} is an adjugate matrix. Based on symmetry, $A_{11} = A_{44}$, $A_{12} = A_{43}$, $A_{13} = A_{42}$, $A_{14} = A_{41}$, $A_{21} = A_{34}$, $A_{22} = A_{33}$, $A_{23} = A_{32}$, and $A_{24} = A_{31}$. Therefore, (4.41) becomes

$$\mathbf{W} = \frac{1}{det(\mathbf{T}_s)} \begin{bmatrix} A_{11} + A_{24} & A_{14} + A_{21} \\ A_{12} + A_{23} & A_{13} + A_{22} \\ A_{13} + A_{22} & A_{12} + A_{23} \\ A_{14} + A_{21} & A_{11} + A_{24} \end{bmatrix} = \begin{bmatrix} W_1 & W_3 \\ W_2 & W_4 \\ W_4 & W_2 \\ W_3 & W_1 \end{bmatrix} \tag{4.42}$$

From the above equation, (4.27) can be rearranged as follows.

$$\mathbf{E} = \mathbf{T}_s \cdot \mathbf{W} \cdot \mathbf{X}$$
$$= \begin{bmatrix} T_1 X_L + T_4 X_R & T_2 X_L + T_3 X_R & T_4 X_L + T_1 X_R & T_3 X_L + T_2 X_R \\ T_2 X_L + T_8 X_R & T_1 X_L + T_7 X_R & T_8 X_L + T_2 X_R & T_7 X_L + T_1 X_R \\ T_8 X_L + T_2 X_R & T_7 X_L + T_1 X_R & T_2 X_L + T_8 X_R & T_1 X_L + T_7 X_R \\ T_4 X_L + T_1 X_R & T_3 X_L + T_2 X_R & T_1 X_L + T_4 X_R & T_2 X_L + T_3 X_R \end{bmatrix}$$
$$\cdot \begin{bmatrix} W_1 \\ W_2 \\ W_3 \\ W_4 \end{bmatrix} \cdot \begin{bmatrix} X_L \\ X_R \\ X_L \\ X_R \end{bmatrix} \tag{4.43}$$

Figure 4.10 shows a block diagram for generating weight vectors using LMS algorithms with the symmetry property.

The weight vectors of the crosstalk canceller are updated based on the LMS adaptive algorithm

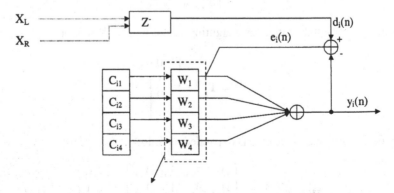

Fig. 4.10. LMS block diagrams for the estimation of crosstalk canceller weighting vectors for the symmetric case.

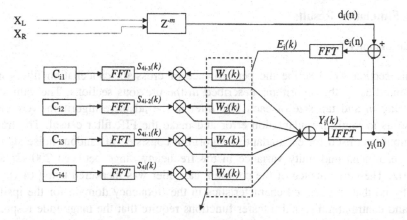

Fig. 4.11. Frequency domain adaptive LMS inverse algorithm block diagrams using overlap-save method for the estimation of crosstalk canceller weighting vectors based on Fig. 4.10 ($i = 1, \ldots, 4$).

$$W_i(n + 1) = W_i(n) + \mu \times (-\hat{\nabla}_i(n)) \qquad i = 1, \ldots, 4 \qquad (4.44)$$

In (4.44), the convergence rate of the LMS adaptive algorithm is controlled by step size μ. The gradient estimate, $\hat{\nabla}(n)$, is simply the derivative of $e^2(n)$ with respect to $W(n)$. Therefore gradient estimates in the time domain can be written as

$$\hat{\nabla}_i(n) = -2[e_1(n)C_{1i}(n) + e_2(n)C_{2i}(n) + e_3(n)C_{3i}(n) + e_4(n)C_{4i}(n)]$$
$$i = 1, \ldots, 4 \qquad (4.45)$$

in which C_{ij} is an element of the ith row and jth column in matrix $\mathbf{C}$

$$\mathbf{C} = \begin{bmatrix} (T_1X_L + T_4X_R) & (T_2X_L + T_3X_R) & (T_4X_L + T_1X_R) & (T_3X_L + T_2X_R) \\ (T_2X_L + T_8X_R) & (T_1X_L + T_7X_R) & (T_8X_L + T_2X_R) & (T_7X_L + T_1X_R) \\ (T_8X_L + T_2X_R) & (T_7X_L + T_1X_R) & (T_2X_L + T_8X_R) & (T_1X_L + T_7X_R) \\ (T_4X_L + T_1X_R) & (T_3X_L + T_2X_R) & (T_1X_L + T_4X_R) & (T_2X_L + T_3X_R) \end{bmatrix}$$
$$(4.46)$$

In (4.43), all input binaural signals and transfer functions are sample sequences in the time domain. The output error is shown in Fig. 4.10. Figure 4.11 shows the block diagram of the frequency domain adaptive LMS inverse algorithm for the symmetric case.

By using the weight vector adaptation algorithm in (4.24) we find

$$W_i(k + 1)$$

$$= W_i(k) + \mu \times F \left\{ F^{-1} \left\{ \frac{S_i^H(k)}{P_i(k)} \cdot E_1(k) + \frac{S_{4+i}^H(k)}{P_{4+i}(k)} \cdot E_2(k) \right. \right.$$

$$\left. \left. + \frac{S_{8+i}^H(k)}{P_{8+i}(k)} \cdot E_3(k) + \frac{S_{12+i}^H(k)}{P_{12+i}(k)} \cdot E_4(k) \right\} \right\}$$

$$i = 1, \ldots, 4 \qquad (4.47)$$

4.3.3 Simulation Results

Single Listener Case

In this section we describe the performance of crosstalk cancellation filters implemented using the algorithms described in the previous sections. The values of the delay m and tap size of each FIR filter in the adaptive algorithm were chosen so as to minimize adaptation error and make the FIR filter causal. The training input data used for each adaptive algorithm consisted of random noise signals with zero mean and unity variance in the frequency range between 200 Hz and 10 kHz. The performance of the crosstalk canceller was measured using (4.18). It was found that the desired characteristics in the frequency domain for the ipsilateral and contralateral signal transfer functions require that the magnitude response of the ipsilateral signal transfer functions in the frequency domain should satisfy $|T_1(\omega)W_1(\omega) + T_2(\omega)W_2(\omega)| = 1$, and $|T_3(\omega)W_3(\omega) + T_4(\omega)W_4(\omega)| = 1$ for lossless signal transfer in the expected frequency band. The ipsilateral signal transfer function should be linear phase: $\angle(T_1(\omega)W_1(\omega) + T_2(\omega)W_2(\omega)) = exp(-jn\omega)$, and $\angle(T_3(\omega)W_3(\omega) + T_4(\omega)W_4(\omega)) = exp(-jn\omega)$. The magnitude response of the contralateral signal transfer functions should satisfy $|T_1(\omega)W_3(\omega) + T_2(\omega)W_4(\omega)| = 0$, and $|T_3(\omega)W_1(\omega) + T_4(\omega)W_2(\omega)| = 0$ for perfect crosstalk cancellation. All of the requirements described above apply to the frequency range between 200 Hz and 10 kHz.

Figure 4.12 shows some typical results for the LMS adaptive inverse algorithm.

The magnitude response of the ipsilateral signal is about 0 dB in the frequency range between 200 Hz and 10 kHz with linear phase. Therefore the desired signal can be transferred from loudspeaker to ear without distortion. The magnitude response

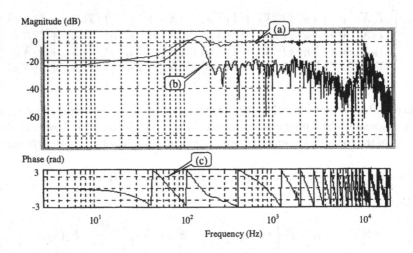

Fig. 4.12. Frequency response of crosstalk canceller adapted in the time domain. (a) Magnitude response of $T_1(\omega)W_1(\omega) + T_2(\omega)W_2(\omega)$; (b) magnitude response of $T_1(\omega)W_3(\omega) + T_2(\omega)W_4(\omega)$; (c) phase response of $T_1(\omega)W_1(\omega) + T_2(\omega)W_2(\omega)$.

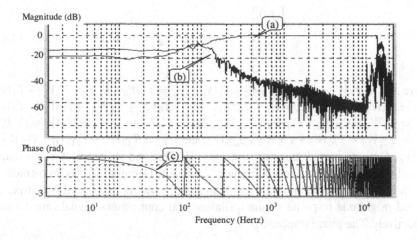

Fig. 4.13. Frequency response of crosstalk canceller adapted in the frequency domain. (a) Magnitude response of $T_1(\omega)W_1(\omega) + T_2(\omega)W_2(\omega)$; (b) magnitude response of $T_1(\omega)W_3(\omega) + T_2(\omega)W_4(\omega)$; (c) phase response of $T_1(\omega)W_1(\omega) + T_2(\omega)W_2(\omega)$.

of the contralateral signal is at least 20 dB below the ipsilateral signal in the same range. Figure 4.13 presents the result of the normalized frequency domain adaptive filter inverse algorithm.

The magnitude response of the ipsilateral signal is about 0 dB in the frequency range between 200 Hz and 10 kHz with linear phase. It has almost the same magnitude response as Fig. 4.12. However, the magnitude response of the contralateral signal is suppressed more than 40 dB below the ipsilateral signal.

Multiple Listener Case

The experiments in this case were conducted as shown in Fig. 4.5. The tap size of the measured HRTFs was 256 samples at a sampling rate of 44.1 kHz. A random noise signal was used for the input of the adaptive LMS algorithm. This signal was sampled at 44.1 kHz in the frequency bands between 200 Hz and 10 kHz. The filter coefficients were obtained using LMS in the time and frequency domain as described above. The performance of the rendering filter for the general nonsymmetric case was measured based on the matrix equation (4.18). The desired magnitude and phase response in the frequency domain should satisfy

$$|M| = \begin{bmatrix} |A_{11}| & |A_{12}| \\ |A_{21}| & |A_{22}| \\ |A_{31}| & |A_{32}| \\ |A_{41}| & |A_{42}| \end{bmatrix}$$

$$= \begin{bmatrix} 1 & 0 \\ 0 & 1 \\ 1 & 0 \\ 0 & 1 \end{bmatrix} \quad 200 \text{ Hz} \leq f \leq 10 \text{ kHz} \quad\quad (4.48)$$

where $A_{11} = T_1W_1 + T_2W_2 + T_3W_3 + T_4W_4$, $A_{12} = T_1W_5 + T_2W_6 + T_3W_7 + T_4W_8$, $A_{21} = T_5W_1 + T_6W_2 + T_7W_3 + T_8W_4$, $A_{22} = T_5W_5 + T_6W_6 + T_7W_7 + T_8W_8$, $A_{31} = T_9W_1 + T_{10}W_2 + T_{11}W_3 + T_{12}W_4$, $A_{32} = T_9W_5 + T_{10}W_6 + T_{11}W_7 + T_{12}W_8$, $A_{41} = T_{13}W_1 + T_{14}W_2 + T_{15}W_3 + T_{16}W_4$, and $A_{42} = T_{13}W_5 + T_{14}W_6 + T_{15}W_7 + T_{16}W_8$. Furthermore, all transfer functions (HRTFs) and weight vectors (crosstalk canceller coefficients) are in the frequency domain. Defining M_{ij} as an element at the ith row and jth column in the above magnitude matrix of $\mathbf{M}$. In this matrix, the desired magnitude response of the ipsilateral and contralateral signals are 1 and 0 respectively. The phase response is

$$\angle(\mathbf{P}) = \begin{bmatrix} \angle A_{11} & \angle A_{12} \\ \angle A_{21} & \angle A_{22} \\ \angle A_{31} & \angle A_{32} \\ \angle A_{41} & \angle A_{42} \end{bmatrix}$$

$$= \begin{bmatrix} e^{-jnw} & X \\ X & e^{-jnw} \\ e^{-jnw} & X \\ X & e^{-jnw} \end{bmatrix} \quad 200 \text{ Hz} \leq f \leq 10 \text{ kHz} \quad\quad (4.49)$$

in which X in the matrix indicates "don't care" because of its small magnitude response. For optimum performance, the ipsilateral signals in (4.49) should have linear phase in the frequency band between 200 Hz and 10 kHz so that there is no phase distortion. Let's define P_{ij} as an element at the ith row and jth column in the phase matrix $\mathbf{P}$. Simulations of (4.48) using LMS adaptation in the time domain are shown in Fig. 4.14 in the frequency domain.

It can be seen that the frequency response of the ipsilateral signal in equation (4.48) is very close to 0 dB in the frequency range between 200 Hz and 10 kHz with linear phase. Therefore the desired ipsilateral signal (input binaural signal) reaches the ear from the same-side loudspeaker without distortion as desired. The magnitude response of the undesired contralateral signal is suppressed between 20 dB and 40 dB relative to the ipsilateral signal in the same frequency range. The same results are shown in the frequency domain in Fig. 4.15.

The frequency response of the ipsilateral signal is nearly identical to the response in Fig. 4.14. The magnitude response of the contralateral signal is suppressed around 40 dB relative to the ipsilateral signal in the same frequency range.

4.3.4 Summary

We described a general methodology for rendering binaural sound over loudspeakers that can be generalized to multiple listeners. We presented the mathematical formulation for the necessary processing of signals to be played from the loudspeakers for

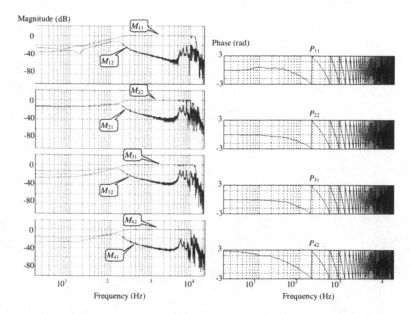

Fig. 4.14. Frequency response where the weight vectors were obtained based on the LMS algorithm in the *time domain*.

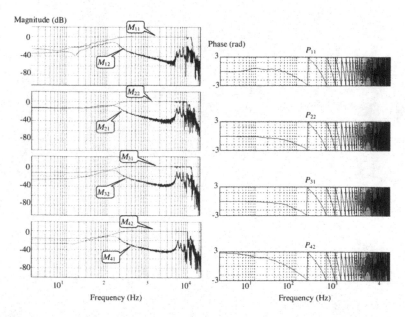

Fig. 4.15. Frequency response where the weight vectors were obtained based on the LMS algorithm in the *frequency domain*.

the case of two listeners. The methods presented here can also be scaled to more listeners.

5

Multiple Position Room Response Equalization

This chapter is concerned with the equalization of acoustical responses, simultaneously, at multiple locations in a room. The importance of equalization is well known, in that it allows (i) delivery of high-quality audio delivered to listeners in a room, and (ii) improved rendering of spatial audio effects for a sense of audio immersion. Typical applications include home theater, movie theaters, automobiles, and any loudspeaker based playback environment (headphones, cell phones, etc.). Because experiencing movies and music is now primarily a group experience (such as in home theaters, automobiles, and movie theaters), and headphone/earbud acoustics vary due to ear coupling effects, it is important to include acoustic variations in the design of an equalization filter. Thus, an equalization filter designed to compensate for the room effects (viz., multipath reflections) at a single location performs poorly at other locations in a room. This is because room impulse responses vary significantly with differing source receiver (viz., listener) positions. A good equalization filter should compensate the effects of multipath reflections simultaneously over multiple locations in a room. This chapter briefly introduces some traditional room equalization techniques, and presents in detail a new multiple listener (or multiple position) equalization filter using pattern recognition techniques. Because the filter lengths can be large, a popular psychoacoustic scheme described in this chapter allows design of low filter orders, using the pattern recognition technique, for real-time implementation. Additionally, a room response and equalization visualization technique, the Sammon map, is presented to interpret the results. Furthermore, one of the major factors that affects equalization performance is the reverberation of the room. In this chapter, the equalization performance of the pattern recognition method [60] is compared with the well-known root mean square averaging-based equalization, using the image method [61] for synthesizing responses with varying reverberation times T_{60}.

5.1 Introduction

An acoustic enclosure can be modeled as a linear system whose behavior is characterized by a response, known as the impulse response, $h(n); n \in \{0, 1, 2, \dots\}$. When the enclosure is a room the impulse response is known as the room impulse response with a frequency response, $H(e^{j\omega})$. Generally, $H(e^{j\omega})$ is also referred to as the room transfer function (RTF). The impulse response yields a complete description of the changes a sound signal undergoes when it travels from a source to a receiver (microphone/listener) via a direct path and multipath reflections due to the presence of reflecting walls and objects. By its very definition the room impulse response is obtained at a receiver (e.g., a microphone) located at a predetermined position in a room, after the room is excited by a broadband source signal such as the MLS or the logarithmic chirp signal (described in chapter 3).

It is well established that room responses change with source and receiver locations in a room [11, 63]. Other reasons for minor variations in the room responses are due to changes in the room, such as opening/closing of doors and windows. When these minor variations are ignored, a room response can be uniquely defined by a set of spatial coordinates, $l_i \triangleq (x_i, y_i, z_i)$. It is assumed that the source is at an origin and the the receiver i is at the three spatial coordinates, x_i, y_i, and z_i, relative to a source in the room.

When an audio signal is transmitted in a room, the signal is distorted by the presence of reflecting boundaries. One scheme to minimize this distortion, from a source to a specific position, is to introduce an *equalizing* filter that is an inverse of the room impulse response measured between the source and the listening position. This equalizing filter is applied to the source signal before transmitting it in a room. If $h_{eq}(n)$ is the equalizing filter for room response $h(n)$, then, for perfect equalization $h_{eq}(n) \otimes h(n) = \delta(n)$; where $\otimes$ is the convolution operator and $\delta(n) = 1, n = 0; 0, n \neq 0$ is the Kronecker delta function. However, two problems arise due to this approach: (i) the room response is not necessarily invertible (i.e., it is not minimum phase), and (ii) designing an equalizing filter for a specific position will introduce a poor equalization performance at other positions in a room. In other words, the multiple point equalization cannot be achieved by an equalizing filter that is designed for equalizing the response at only one location.

A classic multiple location equalization technique is to average the room responses and invert the resulting minimum-phase part to form the equalizing filter. Elliott and Nelson [64] propose a least squares method for designing an equalization filter for a sound reproduction system by adjusting the filter coefficients to minimize the sum of the squares of the errors between the equalized signals at multiple points in a room and the delayed version of an electrical signal applied to a loudspeaker. In [65], Mourjopoulos proposes a technique of using a spatial equalization library, based on the position of a listener, for equalizing the response at the listener position. The library is formed via vector quantization of room responses. Miyoshi and Kaneda [67] present an "exact" equalization of multiple point room responses. Their argument is based on the MINT (multiple-input/multiple-output inverse theorem) which requires that the multiple room responses have uncommon zeros among them. A multiple point equalization algorithm using common acoustical poles is demonstrated by Haneda et al. [68]. Fundamentally, the aforementioned multiple point equalization algorithms are based on a linear least squares approach. Weiss et al. [62] proposed an efficient and effective multirate signal processing-based approach for performing equalization at the listeners' ear positions.

Thus, the main objective of room equalization is the formation of an inverse filter, $h_{eq}(n)$, that compensates for the effects of the loudspeaker and room that cause sound quality degradation at a listener position. In other words, the goal is to satisfy $h_{eq}(n) \otimes h(n) = \delta(n)$, where $\otimes$ denotes the convolution operator and $\delta(n)$ is the Kronecker delta function. Because it is well established that room responses change with source (i.e., loudspeaker) and listener locations in a room [11, 63], clearly, due to the variations in the impulse responses, between positions, equalization has to be done simultaneously such that the goal is satisfied at all listening positions. In practice an ideal delta function is not achievable with low filter orders as room responses are nonminimum-phase. Furthermore, from a psychoacoustic standpoint, a target curve, such as a low-pass filter having a reasonably high cutoff frequency is generally applied to the equalization filter (and hence the equalized response) to prevent the played-back audio from sounding exceedingly "bright". An example of a low-pass cutoff frequency is the frequency where the loudspeaker begins its high-frequency roll-off in the magnitude response. Additionally, the target curve may also be customized according to the size and/or the reverberation time of the room. A high-pass filter may also be applied to the equalized response, depending on the loudspeaker size and characteristics (e.g., a satellite channel loudspeaker), in order to minimize distortions at low frequencies. Examples of environments where multiple listener room response equalization is used are in home theater (e.g., a multichannel 5.1 system), automobile, movie theaters, and the like.

5.2 Background

To understand the effects of single location equalization on other locations, consider a simple first-order *specular* room reflection model as follows (with the assumption that the response at the desired location for equalization is invertible). Let the impulse responses, $h_1(n)$ and $h_2(n)$, from a source to two positions 1 and 2 be represented as

$$h_1(n) = \delta(n) + \alpha_2\delta(n-1); \quad |\alpha_2| < 1$$
$$h_2(n) = \delta(n) + \beta_2\delta(n-1)$$
$$\beta_2 \neq \alpha_2 \tag{5.1}$$

This first-order reflection model is valid, for example, when the two positions are located along the same radius from a source, and each position has a differently absorbing neighboring wall with negligible higher-order reflections from each wall. For simplicity, the absorption due to air and the propagation delay n_d in samples ($n_d \approx f_s r/c$; r is the distance, f_s is the sampling rate and c is the speed of sound which is temperature dependent) is ignored in this model. Ideal equalization at position 1 is achieved if the equalizing filter, $h_{eq}(n)$, is

$$h_{eq}(n) = (-\alpha_2)^n u(n) \qquad (5.2)$$

because $h_{eq}(n) \otimes h_1(n) = \delta(n)$. However, the equalized response at position 2 can be easily shown to be

$$h_{eq}(n) \otimes h_2(n) = \delta(n) - (\alpha_2 - \beta_2)(-\alpha_2)^{(n-1)} u(n-1) \qquad (5.3)$$

where $u(n) = 1, n \geq 0$ is the discrete step function. There are two objective measures of equalization performance for position 2: (i) frequency domain error function (used subsequently in the chapter), and (ii) time domain error function. The time domain error function is easy to compute for the present problem, and is defined as

$$\epsilon = \frac{1}{I} \sum_{n=0}^{I-1} e^2(n) = \frac{1}{I} \sum_{n=0}^{I-1} (\delta(n) - h_{eq(n)} \otimes h_2(n))^2$$

$$= \frac{(\alpha_2 - \beta_2)^2}{I} \sum_{n=1}^{I-1} (-\alpha_2)^{(2n-2)} \qquad (5.4)$$

Clearly, the response at position 2 is unequalized because $\epsilon > 0$. A plot of the error ϵ as a function of the distance $|\alpha_2 - \beta_2|$ between the two coefficients, α_2 and β_2, that differentiate the two responses is shown in Fig. 5.1. Hence, the error is reduced at position 2 if a good equalizer is designed that accounts for the changes in the room response due to variations in the source and listening positions.

5.3 Single-Point Room Response Equalization

Neely and Allen [69] discuss a method of using a minimum-phase inverse filter, wherein the minimum-phase inverse filter is obtained by inverting the causal and stable minimum-phase part of the room impulse response. If the room response is minimum-phase, then perfect (flat) equalization is achieved. However, if the room response is nonminimum-phase, then the resulting minimum-phase inverse filter produces an equalization that contains an audible tone (with a test speech signal). The paper also proposes a method for determining whether the room response is minimum-phase or nonminimum-phase. They state that a *necessary and sufficient condition* for determining whether the room response is nonminimum-phase is through the Nyquist criterion for determining the nonminimum-phase zeros.

Radlović and Kennedy in [70], propose a minimum-phase and all-pass decomposition of a room impulse response. The minimum-phase component is obtained by a combined means of homomorphic processing (cepstrum analysis) and an iterative algorithm. The authors argue that equalizing the phase response in a mixed phase room impulse response is important along with the magnitude equalization, because phase distortions can audibly degrade speech. By using the concept of matched filtering they are able to objectively minimize the total equalization error (magnitude and phase).

Some recent literature on spectral modeling using psychoacoustically motivated filters for single-position equalization can be found in [71, 72, 73].

5.4 Multiple-Point (Position) Room Response Equalization

Miyoshi and Kaneda [67] present an "exact" equalization of multiple-point room responses. Their argument is based on the MINT (multiple-input/multiple-output inverse theorem) which requires that the multiple room responses have uncommon zeros among them. Clearly this is a limiting approach, as uncommon zeros between room responses cannot be guaranteed. The authors take precautions for avoiding common zeros between room responses in their experiments. This is done by avoiding symmetrical positions of the microphones and loudspeakers that are used to measure the room responses in a room.

Chang [74] proposes an universal approximator, such as a neural network, for equalizing the combined loudspeaker and room response system. The authors suggest that the loudspeaker's nonlinear transfer function necessitates the use of a nonlinear inverse filter (neural network). The authors use a time plot to show the low equalization error obtained on using a neural network.

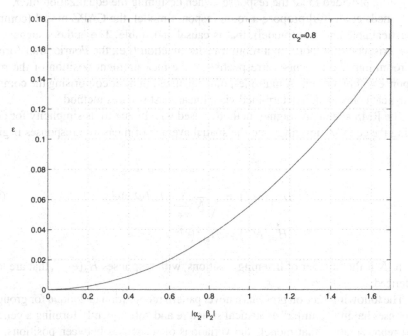

Fig. 5.1. Equalization error at position 2 as a function of the "separation" between the responses at position 1 and 2.

A technique of inverting a mixed-phase response via a least squares approach is presented in [75]. Ideally an inverted version of the room response convolved with the response should provide a Kronecker delta function. However, because room responses are mixed-phase, the room response can be decomposed into a stable causal part, and a noncausal part. Because this noncausal part is unavailable for real-time computation, the author proposes a Levinson algorithm for determining an optimal inverse filter. The objective function for minimization in the Levinson algorithm is

$$J = \sum_{k=0}^{M-p} (\delta(k-p) - y(k))^2 \tag{5.5}$$

where p is the modeling delay, M is the duration of the room response $h(k)$, and $y(k) = h(k) \otimes h_i(k)$ ($\otimes$ denotes the linear convolution operator, and $h_i(k)$ is the causal and finite duration inverse filter of $h(k)$).

Elliott and Nelson [64] propose a method for designing an equalization filter for a sound reproduction system by adjusting the filter coefficients to minimize the sum of the squares of the errors between the equalized responses at multiple points in a room and the delayed version of an electrical signal. Basically, the objective function is expressed as a square of the instantaneous error signal, where the error signal is the difference between a delayed replica of the electrical signal which is supplied as an input to a channel with given room response, and an output signal. The disadvantage of this approach is the relatively limited equalization performance due to the equal weighting provided to all the responses when designing the equalization filter.

Haneda et al. [68] propose a room response model, the CAPZ model (common acoustical pole and zero model), that is causal and stable. The authors suggest that there exist common poles in a room transfer function (i.e., the Fourier transform of the room impulse response) irrespective of the measurement position of the room response within a room. A multiple-point equalization filter comprising the common acoustical poles is then determined via a linear least squares method.

The RMS spatial averaging method is used widely due to its simplicity for computing the equalization filter and the spatial average of measured responses is given by:

$$H_{\text{avg}}(e^{j\omega}) = \sqrt{\frac{1}{N} \sum_{i=1}^{N} |H_i(e^{j\omega})|^2} \tag{5.6}$$

$$H_{\text{eq}}(e^{j\omega}) = H_{\text{avg}}^{-1}(e^{j\omega})$$

where N is the number of listening positions, with responses $H_i(e^{j\omega})$, that are to be equalized.

The following section presents a novel pattern recognition technique for grouping responses having "similar" acoustical structure and subsequently forming a generalized representation that models the variations of responses between positions. The generalized representation is then used for designing an equalization filter.

5.5 Designing Equalizing Filters Using Pattern Recognition

5.5.1 Review of Cluster Analysis in Relation to Acoustical Room Responses

From a broad perspective, clustering algorithms group data that have a high degree of similarity into classes or clusters having centroids. Clustering techniques typically use an objective function, such as a sum of squared distances from the centroids, and seek a grouping (viz., cluster formation) that extremizes this objective function [76]. In particular, clustering refers to assigning data, such as room responses $\{h_i(n); i = 1, 2, \ldots, M; n = 0, 1, \ldots, N - 1\}$, from a data universe X^d comprising a collection of room responses, such that X^d is optimally or suboptimally partitioned into c clusters ($1 < c < M, M < N$). The trivial case of $c = 1$ denotes a rejection of the hypothesis that there are clusters in the data comprising the room responses, whereas $c = M$ constitutes the case where each data vector $\underline{h}_i \triangleq (h_i(0), h_i(1), \ldots, h_i(N - 1))^T$ is in a cluster by itself. Upon clustering, the room responses bearing strong *similarity* to each other are grouped in the same cluster. The similarity between the room responses is used to determine the cluster centroids, and these centroids are then used as a model for the data in their respective clusters. A similarity measure widely used in clustering is the Euclidean distance between pairs of room responses. If the clustering algorithm yields clusters that are well formed, the Euclidean distance between data in the same cluster is significantly less than the distance between data in different clusters.

5.5.2 Fuzzy c-means for Determining the Prototype

In the hard c-means clustering algorithm, a room response, $\underline{h}_j$, can strictly belong to one and only one cluster. This is accomplished by the binary membership function $\mu_i(\underline{h}_j) \in \{0, 1\}$ which indicates the presence or absence of the response $\underline{h}_j$ within a cluster i.

However, in fuzzy clustering, a room response $\underline{h}_j$ may belong to more than one cluster by different "degrees". This is accomplished by a continuous membership function $\mu_i(\underline{h}_j) \in [0, 1]$. The motivation for using the fuzzy c-means clustering approach can be best understood from Fig. 5.2, where the direct path component of the response associated with position 3 is similar (in the Euclidean sense) to the direct path component of the response associated with position 1 (because positions 1 and 3 are at same radial distance from the loudspeaker). Furthermore, it is likely that the reflection components at the position 3 response will be similar to the reflection components of the position 2 response (due to the proximity of these two positions relative to each other). Thus, it is clear that if responses at positions 1 and 2 are clustered separately into two different clusters, then the response at position 3 should belong to both clusters to some degree. Thus, this clustering approach permits an intuitively reasonable model for centroid formation.

The centroids and membership functions, as given in [77, 76], are determined by

$$\underline{\hat{h}}_i^* = \frac{\sum_{k=1}^{M} (\mu_i(\underline{h}_k))^2 \underline{h}_k}{\sum_{k=1}^{M} (\mu_i(\underline{h}_k))^2}$$

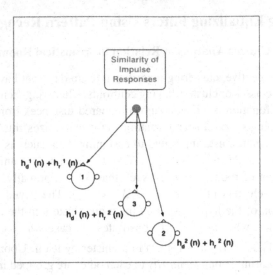

Fig. 5.2. Motivation for using fuzzy c-means clustering for room acoustic equalization.

$$\mu_i(\underline{h}_k) = \left[\sum_{j=1}^{c}\left(\frac{d_{ik}^2}{d_{jk}^2}\right)\right]^{-1} = \frac{\frac{1}{d_{ik}^2}}{\sum_{j=1}^{c}\frac{1}{d_{jk}^2}};$$

$$d_{ik}^2 = \|\underline{h}_k - \hat{\underline{h}}_i^*\|^2 \qquad i = 1, 2, \ldots, c; \; k = 1, 2, \ldots, M \qquad (5.7)$$

where $\hat{\underline{h}}_i^*$ denotes the ith cluster room response centroid. An iterative optimization procedure proposed by Bezdek [76] was used for determining the quantities in (5.7). Care was taken to ensure that the minimum phase room responses were used to form the centroids so as to avoid undesirable time and frequency domain effects, due to incoherent linear combination, resulting from using the excess phase parts.

Once the centroids are formed from minimum-phase responses, they are combined to form a single final prototype. One approach to do this is by using the following model,

$$\underline{h}_{final} = \frac{\sum_{j=1}^{c}(\sum_{k=1}^{M}(\mu_j(\underline{h}_k))^2)\hat{\underline{h}}_j^*}{\sum_{j=1}^{c}(\sum_{k=1}^{M}(\mu_j(h_k))^2)} \qquad (5.8)$$

The final prototype (5.8) is formed from a nonuniform weighting of the cluster membership functions. Specifically, the "heavier" the weight of a cluster j, in terms of the fuzzy membership functions $\sum_{k=1}^{M}(\mu_j(\underline{h}_k))^2$, the larger is the contribution of the corresponding centroid $\hat{\underline{h}}_j^*$ in the formation of the prototype and the subsequent multiple position equalization filter.

The multiple listener equalization filter can subsequently be obtained by determining the minimum-phase component, $\underline{h}_{min,final}$, of the final prototype $\underline{h}_{final} = \underline{h}_{min,final} \otimes \underline{h}_{ap,final}$ ($\underline{h}_{ap,final}$ is the all-pass component), where the minimum-phase sequence $\underline{h}_{min,final}$ is obtained from the cepstrum of $\underline{h}_{final}$. It is noted that the final prototype, $\underline{h}_{final}$, need not be of minimum-phase because the linear combination of minimum-phase signals need not be minimum-phase.

5.5.3 Cluster Validity Index

One approach to determine the optimal number of clusters c^*, based on given data, is to use a validity index κ. One example of a validity index, that is popular in the pattern recognition literature, is the Xie–Beni cluster validity index κ_{XB} [78]. This index is expressed as

$$\kappa_{XB} = \frac{1}{N\beta} \sum_{j=1}^{c} \sum_{k=1}^{M} (\mu_j(\underline{h}_k))^2 \|\hat{\underline{h}}_j^* - \underline{h}_k\|_2^2 \tag{5.9}$$

$$\beta = \min_{i \neq j} \|\hat{\underline{h}}_i^* - \hat{\underline{h}}_j^*\|_2^2$$

The term included with the double summation is simply the objective function used in fuzzy c-means clustering, whereas the denominator term β analyzes the inter-cluster centroid distances. The larger this distance, the better is the cluster separation and thus the lower is the Xie–Beni index.

Thus, the clustering process involves (i) choosing the number of clusters, c, initially to be 2; (ii) performing fuzzy clustering and determining the centroid positions according to Eq. (5.7); (iii) determining κ_{XB} Eq. (5.9); (iv) increasing the number of clusters by unity and performing steps (ii) to (iv) until $c = M - 1$; and (v) plotting κ_{XB} as a function of the number of clusters, where the minima of this plot will provide the optimal number of clusters, c^* according to this index. Typically, κ_{XB} returns the optimal number of clusters $c^* \ll M$ for applications involving very large data sets [79]. In such cases the plot of κ_{XB} versus c increases beyond c^* and then monotonically decreases towards $c = M - 1$. The prototype is then formed from the centroids for c^* via 5.8.

It is to be noted that the equalization filter computed simply by using the fuzzy c-means approach is generally high-order (i.e., at most the length of the room impulse response). Thus, a technique for mapping the large filter lengths to a smaller length is introduced in the subsequent section.

5.5.4 Multiple Listener Room Equalization with Low Filter Orders

Linear predictive coding (LPC) [80, 81] is used widely for modeling speech spectra with a fairly small number of parameters. It can also be used for modeling room responses in order to form low-order equalization filters.

In addition, in order to obtain a better fit of a low-order model to a room response, especially in the low-frequency region of the room response spectrum, the concept of warping was introduced by Oppenheim et al. in [82]. Warping involves the use of a chain of all-pass blocks, $D_1(z)$, instead of conventional delay elements z^{-1}, as shown in Fig. 5.3. With an all-pass filter, $D_1(z)$, the frequency axis is warped and

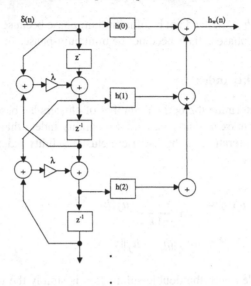

Fig. 5.3. The structure for implementing warping.

the resulting frequency response is obtained at nonuniformly sampled points along the unit circle. Thus, for warping the axis transformation is achieved by

$$D_1(z) = \frac{z^{-1} - \lambda}{1 - \lambda z^{-1}} \tag{5.10}$$

The group delay of $D_1(z)$ is frequency-dependent, so that positive values of the warping coefficient λ yield higher frequency resolutions in the original response for low frequencies, whereas negative values of λ yield higher resolutions for high frequencies. The cascade of all-pass filters results in an infinite duration sequence, hence typically a window is employed that truncates this infinite duration sequence to a finite duration to yield an approximation.

Smith and Abel [83] proposed a bilinear conformal map based on the all-pass transformation (5.10) that achieves a frequency warping nearly identical to the Bark frequency scale (also called the critical band rate) [84, 26]. They found a closed-form expression that related the warping coefficient of the all-pass transformation to the sampling frequencies $f_s \in (1 \text{ kHz}, 54 \text{ kHz}]$ that achieved this psychoacoustically motivated warping transformation. Specifically it was shown that

$$\lambda = 0.8517\sqrt{\arctan(0.06583 f s/1000)} - 0.1916 \tag{5.11}$$

In the subsequent details, it is to be understood that $\lambda = 0.77$ (for an f_s of 48 kHz) was used. The warping induced between two frequency axes by Eq. (5.10) is depicted in Fig. 5.4 for different values of the warping coefficients λ. The frequency

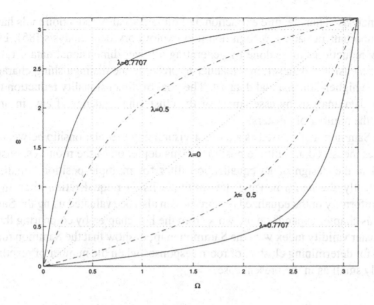

Fig. 5.4. Frequency warping for various values of λ.

resolution plot for different warping coefficients, λ, is shown in Fig. 5.5. It can be seen that the warping to the Bark scale for $\lambda = 0.77$ gives a "balanced" mapping because it provides a good resolution at low frequencies while retaining the resolution at mid and high frequencies (e.g., compare with $\lambda = 0.99$). Some recent literature on spectral modeling using warping can be found in [71, 72, 73].

The general system-level approach for determining the cluster-based multiple listener equalization filter is shown in Fig. 5.6. Specifically, the room responses are initially warped to the psychoacoustical Bark scale. As later shown, the Xie–Beni cluster validity index gives an indication of the number of clusters that are generated for the given data set (particularly for the case where the number of data samples, M, is relatively small). Subsequently, the number of clusters is used for performing clustering in order to determine the cluster centroids and prototype, respectively. The minimum-phase part of the prototype, having length $N \gg 2$, is then parameterized by a low-order model, such as the LPC, for realizable implementation. The inverse filter is then found from the LPC coefficients, and the reverse step of unwarping is performed to obtain the filter in the linear domain. The equalization performance can then be assessed by inspecting the equalized responses along a log frequency domain.

5.6 Visualization of Room Acoustic Responses

Visualizing information generated by a system is important for understanding its characteristics. Frequently, signals in such systems are multidimensional and need to be displayed on a two-dimensional or three-dimensional display for facilitating visual analysis. Thus, feature extraction and dimensionality reduction tools have become important in pattern recognition and exploratory data analysis [85]. Dimensionality reduction can be done by generating a lower-dimensional data set, from a higher-dimensional data set, in a manner to preserve the distinguishing characteristics of the higher-dimensional data set. The goal of dimensionality reduction then is to allow data analysis by assessment of the clustering tendency, if any, in order to identify the number of clusters.

The Sammon map is used as a tool for visualizing the relationship between room responses measured at multiple positions. This depiction of the room responses can also aid in the design of an equalization filter for multiple position equalization. Subsequently, the performance of this multiple-listener equalization filter, in terms of the uniformity of the equalized responses, can also be evaluated using the Sammon map. This chapter expands on the work from the last chapter, by comparing the Xie–Beni cluster validity index with the Sammon map, to show that the Sammon map can be used for determining clusters of room responses when the number of responses is relatively small as in the present case.

5.7 The Sammon Map

In 1969, Sammon [86] introduced a method to map multidimensional data onto lower dimensions (e.g., 2 or 3). The main property of the Sammon map is that it retains the geometrical distances between signals, from multidimensional space, on the two-dimensional or three-dimensional space [87].

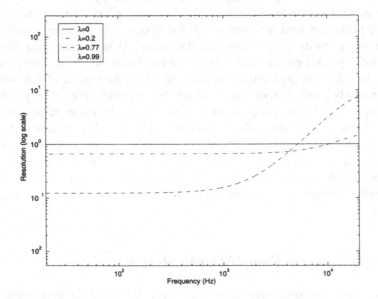

Fig. 5.5. The frequency resolution with different warping coefficients λ.

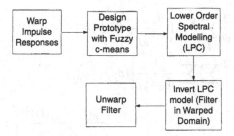

Fig. 5.6. System for determining the the multiple listener equalization filter based on perceptual pattern recognition.

Consider $\{h_i(n), n = 0, 1, \ldots, N - 1\}$ to be the room responses associated with locations $i = 1, \ldots, M$ ($M >= 2$), where each of these responses is of duration $N \gg 2$ samples. Let $d_{ij} = \|h_i(n) - h_j(n)\|_2$ be the Euclidean distance between the room responses at positions i and j, respectively. Let $\{r_i(l)\}, l \in \{0, 1\}$ be the location of the image of $\{h_i(n)\}$ on a two-dimensional display. The goal is to position the $\{r_i(l)\}$ ($i = 1, \ldots, M$) onto the display in such a way that all their mutual Euclidean distances $\|r_i(l) - r_j(l)\|_2$ approximate the corresponding distances d_{ij}. Thus, distances in multidimensional spaces are mapped to approximately equivalent distances in two dimensions via the Sammon mapping.

The objective function, E, that governs the adaptive Sammon map algorithm to converge to a locally optimal solution (where distances are approximated) is given as

$$E = \frac{1}{\sum_{i=1}^{M} \sum_{j>i} d_{ij}} \sum_{i=1}^{M} \sum_{j>i} \frac{(d_{ij} - \|r_i(l) - r_j(l)\|_2)^2}{d_{ij}} \qquad (5.12)$$

Fundamentally, it is desired to adjust the $r_i(l) \in \Re^2$ so as to minimize the objective function E by a gradient descent scheme. Once a locally optimal solution is found, the $r_i(l)$s are configured on a two-dimensional plane such that the relative distances between the different $h_i(n)$ are visually discernible. In this chapter, Sammon mapping for $M = 6$ responses was performed, but the technique can be easily adapted for more responses.

With the following notation,

$$\phi = \sum_{i=1}^{M} \sum_{j>i} d_{ij}$$

$$\underline{r}_p \triangleq (r_p(0), r_p(1))^T \qquad (5.13)$$

$$d'_{pj} = \|\underline{r}_p - \underline{r}_j\|_2$$

the gradient descent algorithm, at iteration m, for determining $\underline{r}_p$ is given by

$$^{(m+1)}r_p(l) = {}^{(m)}r_p(l) - \alpha \frac{\frac{\partial E(m)}{^{(m+1)}\partial r_p(l)}}{\left| \frac{\partial^2 E(m)}{^{(m+1)}\partial r_p^2(l)} \right|}; \qquad l = \{0, 1\} \qquad (5.14)$$

$$\frac{\partial E(m)}{(m+1)\partial r_p l} = -\frac{2}{\phi} \sum_j \sum_{p\neq j} \left(\frac{d_{pj} - d'_{pj}}{d_{pj}d'_{pj}}\right)(r_p(l) - r_j(l))$$

$$\frac{\partial^2 E(m)}{(m+1)\partial r_p^2(l)} = -\frac{2}{\phi} \sum_j \sum_{p\neq j} \frac{1}{d_{pj}d'_{pj}} \left[(d_{pj} - d'_{pj})\right.$$

$$\left. -\frac{(r_p(l) - r_j(l))^2}{d'_{pj}}\left(1 + \frac{d_{pj} - d'_{pj}}{d'_{pj}}\right)\right]$$

where α was set to 0.3.[1]

In essence, the Sammon map, which is a nonlinear projection algorithm belonging to the class of metric multidimensional scaling (MDS) algorithms ([88, 89]), permits visualization of class or cluster distributions of multidimensional data on a 2-D plane. The computational complexity of the map can be fairly high if the number of data points, M, is large because the objective function (1) is based on $O(M^2)$ distances. For small M, as in this chapter, the Sammon mapping imposed negligible computational requirements and results were obtained in less than half a minute on a Pentium IV 2.66 GHz. Speedups of the Sammon algorithm can be found, for example, in [90].

5.8 Results

A listening arrangement is shown in Fig. 5.7, where the microphone locations for measuring the room responses were at the center of the listener head at each position. The distance from the loudspeaker to the listener 2 position was about 7 meters, whereas the average intermicrophone distance in the listener arrangement was about 1 meter. The room was roughly of dimensions 10 m × 20 m × 6 m. The resulting measurement was captured by an omnidirectional flat response microphone and deconvolved by the SYSid system. The chirp was transmitted around 30 times and the resulting measurements were averaged to get a higher SNR. The omnidirectional microphone had a substantially flat magnitude response (viz., the Countryman ISO-MAX B6 Lavalier microphone). The loudspeaker was a center channel speaker from a typical commercially available home theater speaker system.

First, the Sammon map is applied to the six psychoacoustically warped responses for visualizing the responses on a 2-D plane. One of the goals of this step is to see if the map captures any perceptual grouping between the responses. Subsequently the fuzzy c-means clustering algorithm on $M = 6$ is then applied to the psychoacoustically warped room responses (each response being a vector of length 8192) and the optimal number of clusters using the Xie–Beni index (Eq. (5.9) in the previous chapter) is determined. As shown, the Sammon map, when compared to the Xie–Beni cluster validity index, gives a clear indication of the number of clusters that are generated for the given data set.

---

[1] Sammon [86] called this a magic factor and recommended α to be ≈ 0.3 or 0.4.

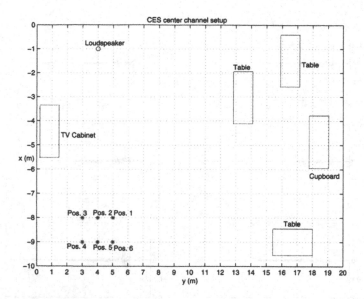

Fig. 5.7. The experimental setup for measuring $M = 6$ acoustical room responses at six positions in a reverberant room.

Figure 5.8 shows the responses at the six listener positions in the time domain. Clearly, there are significant differences in the responses. Firstly, it can be visually observed that there is a certain similarity in the time of arrival of the direct path component of the responses at positions 1, 2, and 3. Also, noticeable is the path delay difference between the responses at positions 4, 5, and 6 in relation to the responses in positions 1, 2, and 3. Figure 5.9 shows the corresponding 1/3 octave smoothed magnitude responses along a linear frequency axis in Hertz. Figure 5.10 shows the corresponding 1/3 octave smoothed magnitude responses in the Bark domain. Specifically, the x-axis (viz., the Bark axis) was computed using the expression provided in [84],

$$z = 13 \tan^{-1}(0.76f/1000) + 3.5 \tan^{-1}(f/7500)^2 \qquad (5.15)$$

where f is the frequency in Hz. A comparison between the plots of Figs. 5.9 and 5.10 shows a transformation effected by the mapping of (5.10) in a sense that low frequencies are mapped higher which effectively "stretches" the magnitude response.

The Sammon map for the warped responses is shown in Fig. 5.11. The map shows the relative proximity of the responses at positions 1, 2, and 3 that could be identified as a group. Table 5.1 below shows all symmetric distances, $d'_{ij} = \|\underline{r}_i - \underline{r}_j\|_2$, between the warped responses i, j, as computed through the Sammon map on the 2-D plane (i.e., the ith row and jth column element is the distance between the Sammon mapped coordinates corresponding to the ith and jth position).

From Fig. 5.11 and Table 5.1, a dominant perceptual grouping of responses 1, 2, and 3 can be seen on the map. This can be confirmed from the distance metrics in Table 5.1 especially where the distance between responses 1, 2, and 3 are significantly

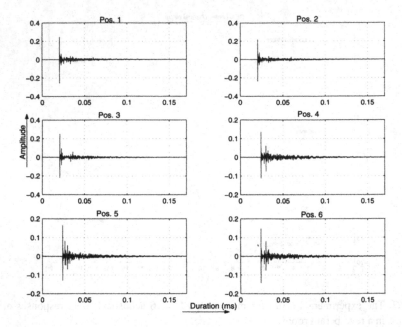

Fig. 5.8. The time domain responses at the six listener positions for the setup of Fig. 5.8.

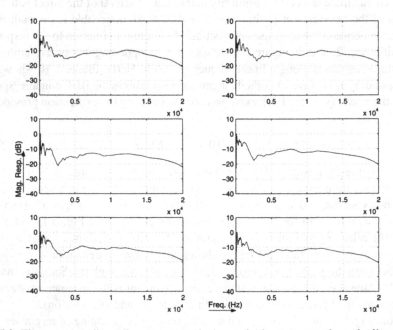

Fig. 5.9. The corresponding 1/3 octave smoothed magnitude responses along the linear frequency axis obtained from Fig. 5.8.

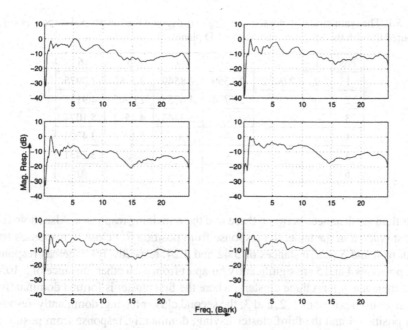

Fig. 5.10. The 1/3 octave smoothed magnitude responses of Fig. 5.9 along the Bark axis.

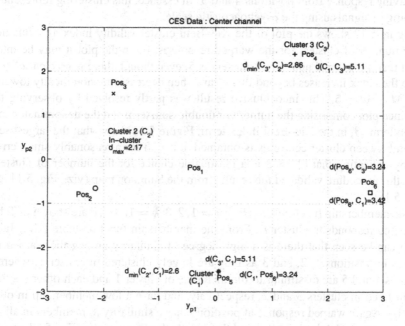

Fig. 5.11. The Sammon map for the warped impulse responses.

Table 5.1. The symmetric distances, $d'_{ij} = \|\underline{r}_i - \underline{r}_j\|_2$, between the warped responses i, j, as computed through the Sammon map on the 2-D plane

Pos	1	2	3	4	5	6
1	0	2.0422	2.1559	2.8549	2.5985	3.0975
2	—	0	2.1774	4.6755	2.9005	4.9133
3	—	—	0	3.3448	4.4522	5.1077
4	—	—	—	0	5.1052	3.4204
5	—	—	—	—	0	3.2431
6	—	—	—	—	—	0

lower than the distances between these and the remaining responses. Also, note from the last column in the table, the response from position 6 is close to responses from positions 4 and 5 (i.e., distances of 3.42 and 3.24, respectively), whereas responses from positions 4 and 5 are significantly far apart from each other (distance of 5.1052). Thus, there are at least three clusters, where the first cluster is formed dominantly of responses from positions 1, 2, and 3; the second cluster having, dominantly, response from position 4 and the third cluster having, dominantly, response from position 5. Finally, response from position 6 could be grouped with at least the distinct clusters having response from positions 4 and 5. In essence, this clustering represents a grouping of signals using the psychoacoustic scale.

Figure 5.12 shows the plot of the Xie–Beni cluster validity index as a function of the number of clusters for the warped responses. From the plot it may be interpreted that the optimal number of clusters is 5, even though there is no clear $c^* \neq 5$ where the index increases beyond this c^* and then decreases monotonically towards $c = M - 1 = 5$. This inconclusive result was partly resolved by observing the combined plot comprising the numerator double-sum term and the denominator separation term, β, in the Xie–Beni index term. Figure 5.13 shows that the largest separation between cluster centroids is obtained at $c = 3$ for a reasonably small error, thereby indicating that $c^* = 3$ is a reasonable choice for the number of clusters. Thus, this procedure validated the results from the Sammon map (viz., Fig. 5.11 and Table 5.1).

The membership functions, $\mu_j(\underline{h}_k), j = 1, 2, 3; k = 1, \ldots, 6$ are shown in Table 5.2 (C_i corresponds to cluster i). From the membership function table (viz., Table 5.2) it can be seen that there is a high degree of similarity among the warped responses in positions 1, 2, and 3 as they are largely clustered in cluster 1, whereas positions 4 and 5 are dissimilar to the members of cluster 1 and each other (as they are clustered in clusters 3 and 2, respectively, and have a low membership in other clusters). Again warped response at position 6 has a similarity to members in all the three clusters (which is also predicted through the Sammon map of Fig. 5.11 and the table relating distances on the Sammon map).

In the next step equalization is performed according to Fig. 5.6, with $c^* = 3$ and LPC order $p = 512$, and depict the equalization performance results using the Sammon map. An inherent goal in this step is to view the equalization performance

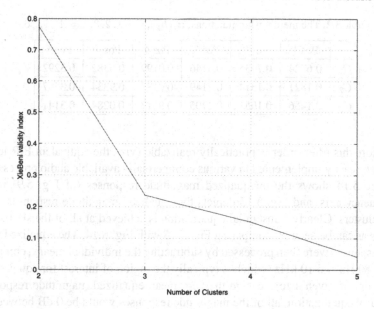

Fig. 5.12. The Xie–Beni cluster validity index as a function of the number of clusters for the warped responses.

results visually, and demonstrate that the uniformity and similarity of the magnitude responses (unequalized and equalized) can be shown on a 2-D plane using the map. The LPC order $p = 512$ was selected as it gave the best results upon equalization and

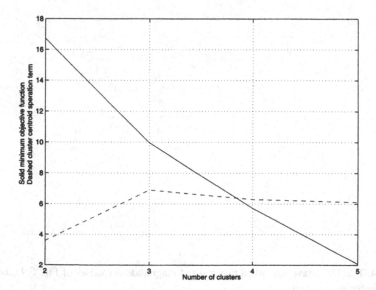

Fig. 5.13. The numerator (viz., the objective function) and denominator term (viz., the separation term) of the Xie–Beni cluster validity index of Fig. 5.12.

Table 5.2. The membership functions, $\mu_j(\underline{h}_k), j = 1, 2, 3; k = 1, \ldots, 6$

	Pos 1	Pos 2	Pos 3	Pos 4	Pos 5	Pos 6
C_1	0.6724	0.7635	0.6146	0.0199	0.0382	0.3292
C_2	0.1821	0.1314	0.2149	0.0185	0.9334	0.3567
C_3	0.1456	0.1051	0.1705	0.9617	0.0284	0.3141

furthermore this filter order is practically realizable (viz., the equivalent FIR length filter can be easily implemented in various commercially available audio processors.)

Figure 5.14 shows the unequalized magnitude responses (of Fig. 5.9) in the log frequency axis, and Fig. 5.15 depicts the equalized magnitude responses using $c^* = 3$ clusters. Clearly, substantial equalization is achieved at all of the six listener positions as can be seen by comparing Fig. 5.15 with Fig. 5.14. The equalized magnitude responses were then processed by subtracting the individual means, computed between 80 Hz and 10 kHz (which is typically the region of interest for equalization in the room of given size), to give the zero mean equalized magnitude responses. Under ideal equalization, all of the magnitude responses would be 0 dB between 80 Hz and 10 kHz. Hence, upon applying the Sammon map, all of the ideal equalized responses would be located at the origin of the Sammon map. Any deviation away from 0 dB would show up on the map as a displacement away from the origin. If the

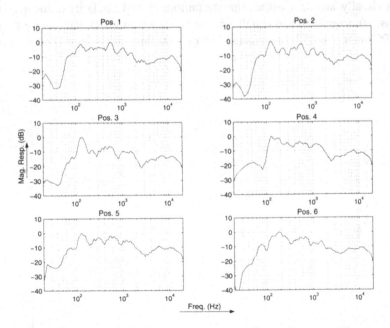

Fig. 5.14. The 1/3 octave smoothed unequalized magnitude responses of Fig. 5.9 (shown in the log frequency domain).

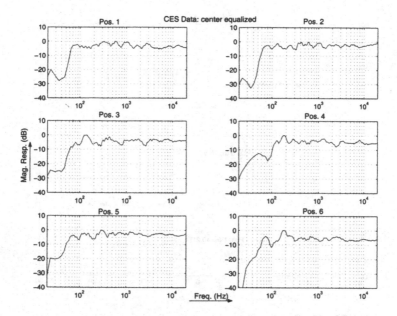

Fig. 5.15. The 1/3 octave smoothed equalized magnitude responses (shown in the log frequency domain for better depiction of performance at low frequencies) using $c^* = 3$ clusters.

equalized responses were uniform in distribution, then they would appear in a tight circle about the origin in the 2-D plane after applying the Sammon map.

Now, applying the Sammon map algorithm to the original magnitude responses of Fig. 5.9, between 80 Hz and 10 kHz, results in Fig. 5.16. Specifically, the responses in a 2-D plane for different positions show significant non-uniformity as these are not located equidistant from the origin. Applying the mean corrected and equalized responses to the Sammon map algorithm gives the distribution of the equalized responses on a 2-D plane as shown in Fig. 5.17.

Comparing Fig. 5.16 with Fig. 5.17 shows an improved uniformity among the responses as many of the responses lie at approximately the same distance from the origin. Specifically, from Fig. 5.17 it is evident that the distances of equalized responses 1, 2, 4, and 5 are close to each other from the origin, thereby reflecting a larger uniformity between these responses. Furthermore, the standard deviation of the distances of the equalized responses is much smaller than that of the unequalized responses (viz., 4.64 as opposed to 11.06) indicating a better similarity between the equalized responses. The improved similarity of the equalized magnitude responses 1, 2, 4, and 5 can be checked by visually comparing the equalized responses in Fig. 5.15. Also, it can be seen that the equalized magnitude responses 3 and 6 are quite a bit different from each other, and from equalized responses 1, 2, 4, and 5, and this reflects in the Sammon map as points 3 and 6 substantially offset from a circular distribution.

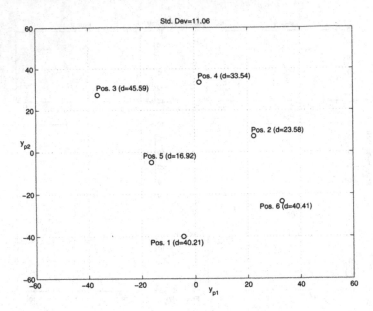

Fig. 5.16. The Sammon map of the unequalized magnitude responses.

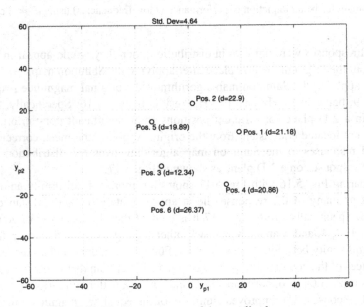

Fig. 5.17. The Sammon map of the equalized responses.

5.9 The Influence of Reverberation on Room Equalization

As is known, room equalization is important for delivering high-quality audio in multiple listener environments and for improving speech recognition rates. Lower-order equalization filters can be designed at perceptually relevant frequencies through warping. However, one of the major factors that affects equalization performance is the reverberation of the room. In this chapter, we compare the equalization performance of the pattern recognition method to the well known root mean square averaging based-equalization using the image method [61].

5.9.1 Image Method

The room impulse response, $p(t, \underline{X}, \underline{X}')$, for the image model [61] with loudspeaker at $\underline{X} = (x, y, z)$ and microphone at $\underline{X}' = (x', y', z')$ and room dimensions $\underline{L} = (L_x, L_y, L_z)$ (with walls having absorption coefficient $\alpha = 1 - \beta^2$) is given as

$$
p(t, \underline{X}, \underline{X}') = \sum_{p=0}^{1} \sum_{r=-\infty}^{\infty} \beta_{x_1}^{|n-q|} \beta_{x_2}^{|n|} \beta_{y_1}^{|l-j|} \beta_{y_2}^{|l|} \beta_{z_1}^{|m-k|} \beta_{z_2}^{|m|} \frac{\delta[t - (|\underline{R}_p + \underline{R}_r|/c)]}{4\pi|\underline{R}_p + \underline{R}_r|}
$$

(5.16)

$$
\begin{aligned}
\underline{p} &= (q, j, k) \\
\underline{R}_p &= (x - x' + 2qx', y - y' + 2jy', z - z' + 2kz') \\
\underline{r} &= (n, l, m) \\
\underline{R}_r &= 2(nL_x, lL_y, mL_z)
\end{aligned}
$$

The room image model (5.16), thus, can be simulated for different reverberation times by adjusting the reflection coefficients (viz., the βs), because the Schroeder reverberation time T_{60} is related to the absorption coefficients by the equation $T_{60} = 0.161V/\sum_i S_i \alpha_i$ (S_i is the surface area of wall i).

Hence, the robustness to reverberation, of different equalization techniques, can be modeled by varying the reflection coefficients in the image model.

5.9.2 RMS Average Filters

The RMS average filter (as used traditionally for movie theater equalization) is obtained as

$$
H_{avg}(e^{j\omega}) = \sqrt{\frac{1}{N} \sum_{i=1}^{N} |H_i(e^{j\omega})|^2}
$$

(5.17)

$$
H_{eq}(e^{j\omega}) = H_{avg}^{-1}(e^{j\omega})
$$

where $|H_i(e^{j\omega})|$ is the magnitude response at position i. To obtain lower-order filters, we used the approach as shown in Fig. 5.6 (but using RMS averaging instead of fuzzy c-means prototype formation).

5.9.3 Results

We have compared the pattern recognition and warping-based method to the RMS averaging and warping-based method, for multiposition equalization, to determine their robustness to reverberation variations. Ideally, it is required that the *equalization performance* does not degrade significantly, when the reverberation time increases, for (i) a fixed room, and (ii) fixed positions of the listeners in a room. The room image model allows ease in simulating changes in responses (due to changes in reverberation times) thereby allowing the equalization performance of these methods to be compared.

To quantify the equalization performance, we used the well-known spectral deviation measure, σ_E, which indicates the degree of flatness of the spectrum. The lower the measure, the better is the performance. The performance measure is defined as

$$
\sigma_E = \sqrt{\left[\frac{1}{P} \sum_{i=0}^{P-1} (10\log_{10} |E(e^{j\omega_i})|) - B(e^{j\omega}) \right]} \tag{5.18}
$$

$$
B(e^{j\omega}) = \frac{1}{P} \sum_{i=0}^{P-1} (10\log_{10} |E(e^{j\omega_i})|)^2
$$

$$
|E(e^{j\omega})| = |H(e^{j\omega})||H_{eq}(e^{j\omega})|
$$

The image model was simulated for a room of volume $L_x \times L_y \times L_z = 8$ m $\times 8$ m $\times 4$ m, the source speaker at $(x, y, z) = (0$ m, 4 m, 1.5 m) and six listeners arranged in a rectangular configuration in front of the source with $y' \in [3$ m, 5 m], $x' \in [3$ m, 4 m], $z' = 1.5$ m. The sampling frequency, f_s, was set at 16 kHz (wideband speech/music equalization in 20 Hz to 8 kHz range).

Figure 5.18 shows the reverberation robustness for the equalization using the pattern recognition method and with 512 FIR taps, whereas Fig. 5.19 shows the reverberation robustness using the RMS averaging method with 512 taps. The equalization performance measure, σ_E, was determined in the 20 Hz to 8 kHz range. The x-axis, in each plot, corresponds to the reverberation time T_{60} used in the simulation, and the y-axis is σ_E.

It can be observed that the pattern recognition method outperforms the RMS averaging method due to lower σ_E for larger T_{60}s at most of the six listener positions. Only four curves can be clearly seen, because the absorption coefficients were kept the same for all walls for a given simulation run. Thus, the symmetry induced by the relative positioning of the source to the microphones delivered the same performance for positions 1 and 3, as well as the same performance for positions 4 and 6.

Also, an interesting observation is that σ_E (an average measure) does not increase monotonically with increasing T_{60} for all listener positions for both methods as some positions are prone to better equalization.

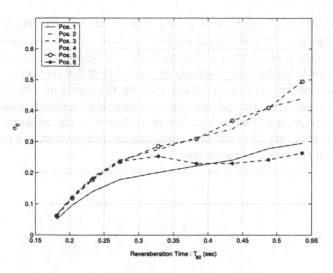

Fig. 5.18. Performance of proposed method (x-axis: T_{60}, y-axis: σ_E).

5.10 Summary

In this chapter some background on various single position and multiple position equalization techniques, including the importance of performing multiple position equalization over single position was presented. Also presented was a pattern recognition method of performing simultaneous multiple listener equalization with low filter orders, and some comparisons between the RMS averaging equalization method and the pattern recognition equalization method in terms of reverberation robustness.

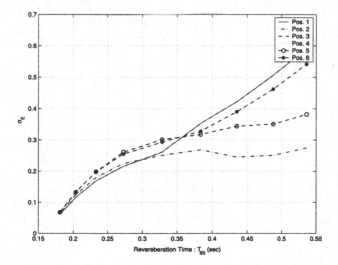

Fig. 5.19. Performance of RMS averaging method (x-axis: T_{60}, y-axis: σ_E).

A technique for visualizing room impulse responses and simultaneous multiple listener equalization performance using the Sammon map was also presented. The map is able to display results obtained through clustering algorithms such as the fuzzy c-means method. Specifically, distances of signals in multidimensional spaces are mapped onto distances in two dimensions, thereby displaying the clustering behavior of the proposed clustering scheme. Upon determining the equalization filter from the final prototype, the resulting equalization performance can be determined from the size and shape (viz., circular shape indicates uniform equalization performance at all locations) of the equalization map.

6

Practical Considerations for Multichannel Equalization

Given a multichannel loudspeaker system, the selection of the crossover frequency between the subwoofer and the satellite speakers is important for accurate (i.e., distortion-free), reproduction of playback sound. Presently, many home theater systems have selectable crossover frequencies, which are part of the bass management filter capabilities, which are set by the consumer through listening tests. Alternatively, if the loudspeakers are industry certified, the crossover frequency is set at 80 Hz. A desirable feature is that, besides distortion-free sound output from the individual subwoofer and the satellite speakers, the combined subwoofer and satellite room acoustical response should exhibit negligible variations around the selected crossover frequency. In this chapter, we present an automatic crossover frequency selection algorithm based on an objective measure (viz., the spectral deviation measure) for multichannel home theater applications that allows better control of the combined subwoofer and satellite response, thereby significantly improving audio quality. Initially, some results are presented that show the effect of crossover frequency on low frequency performance. Additional parameter optimization of the bass management filters is shown to yield improved performance. Comparison between the results from crossover and all-parameter optimization, of the bass management filters, for multiposition equalization is presented. As also shown, cascading an all-pass filter, or adding in time-delays in loudspeaker channels, can provide further improvements to the equalization result in the crossover region. Alternative techniques for fixing the crossover blend, using a cascade of all-pass filters, are also presented.

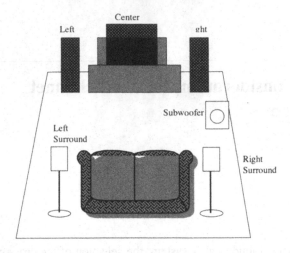

Fig. 6.1. A 5.1 system.

6.1 Introduction

A room is an acoustic enclosure that can be modeled as a linear system whose behavior at a particular listening position is characterized by an impulse response, $h(n); n \in \{0, 1, 2, \dots\}$ with an associated frequency response or room transfer function $H(e^{j\omega})$. The impulse response yields a complete description of the changes a sound signal undergoes when it travels from a source to a receiver (microphone/listener). The signal at a listening position consists of direct path components, discrete reflections that arrive a few milliseconds after the direct sound, as well as a reverberant field component.

A typical 5.1 system and its system-level description are shown in Figs. 6.1 and 6.2, respectively, where the satellites (left, center, right, left surround, and right surround speakers) are positioned surrounding the listener and the subwoofer may be placed in the corner or near the edges of a wall. The high-pass (satellite) and low-pass (subwoofer) bass management filters, $|H^{hp}_{bm,\omega_c}(\omega)| = 1 - 1/\sqrt{1 + (\omega/\omega_c)^4}$ and $|H^{lp}_{bm,\omega_c}(\omega)| = 1/\sqrt{1 + (\omega/\omega_c)^8}$, are Butterworth second-order high-pass (12 dB/octave roll-off) and fourth-order low-pass (24 dB/octave roll-off), respectively, and are designed with a crossover frequency ω_c (i.e., the intersection of the corresponding −3 dB points) corresponding to 80 Hz. Alternatively, the fourth-order Butterworth can be implemented as a cascade of two second-order Butterworth filters which modifies the magnitude response slightly around the crossover region. If the satellite response rolls off at a second-order rate, then the resulting response obtained through complex summation has a flat magnitude response in the crossover region. The analysis and techniques presented in this chapter can be modified in a straightforward manner to include cascaded Butterworth or any other bass management filter. Examples of other crossover networks that split the signal energy between the subwoofer and the satellites, according to predetermined crossover frequency and slopes, can be found in [91, 92, 93]. The magnitude responses of the individual bass management filters as well as the magnitude of the recombined response (i.e., the magnitude of the complex sum of the filter frequency responses), are shown in Fig. 6.3. If the satellite response is smooth and rolling of at a second order Butterworth rate, then the complex summation yields a flat magnitude response in the audio signal pass-band. In real rooms, the resulting magnitude response from the bass management filter set, combined with with the loudspeaker and room responses, will exhibit substantial variations in the crossover region. This effect can be mitigated by proper selection of the crossover frequency (and/or the bass management filter orders). In essence, the bass management filter parameter selection should be such that the subwoofer and the satellite channel output be substantially distortion-free with minimal variations in the crossover region.

The acoustical block diagram for a subwoofer channel and a satellite channel is shown in Fig. 6.4, where $H_{sub}(\omega)$ and $H_{sat}(\omega)$ are the acoustical loudspeaker and room responses at a listening position. The resulting *net* acoustic transfer function, $H(\omega)$, and magnitude response, $|H(e^{j\omega})|^2$, can be written as

$$H(\omega) = H^{hp}_{bm,w_c}(\omega)H_{sat}(\omega) + H^{lp}_{bm,w_c}(\omega)H_{sub}(\omega)$$
$$|H(\omega)|^2 = |A(\omega)|^2 + |B(\omega)|^2 + \Gamma(\omega)$$
$$|A(\omega)|^2 = |H^{hp}_{bm,f_c}(\omega)|^2|H_{sat}(j\omega)|^2$$
$$|B(\omega)|^2 = |H^{lp}_{bm,f_c}(\omega)|^2|H_{sub}(\omega)|^2 \qquad\qquad (6.1)$$
$$\Gamma(\omega) = 2|A||B|\cos(\phi_{sub}(\omega) + \phi^{lp}_{bm,w_c}(\omega) - \phi_{sat}(\omega) - \phi^{hp}_{bm,w_c}(\omega))$$

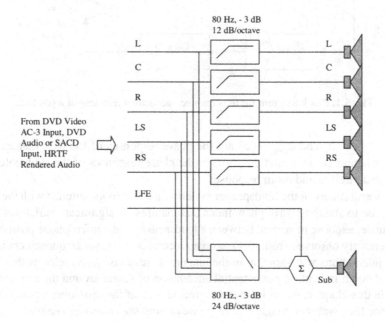

Fig. 6.2. System-level description of the 5.1 multichannel system of Fig. 6.1.

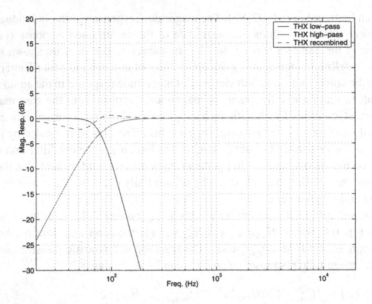

Fig. 6.3. Magnitude response of the industry standard bass management filters and the recombined response.

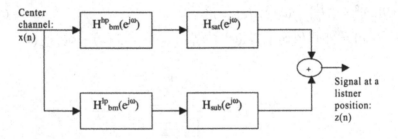

Fig. 6.4. Block diagram for the combined acoustical response at a position.

where $\phi_{bm,\omega_c}^{hp}(\omega)$ and $\phi_{bm,\omega_c}^{lp}(\omega)$ are the phase responses of the bass management filters, whereas $\phi_{sub}(\omega)$ and $\phi_{sat}(\omega)$ are the phase responses of the subwoofer and room, and satellite and room responses.

However, many of the loudspeaker systems, in a real room, interact with the room giving rise to standing wave phenomena that manifest as significant variations in the magnitude response measured between a loudspeaker and a microphone position. As can be readily observed from (6.1), with an incorrect crossover frequency choice the phase interactions will show up in the magnitude response as a region with a broad spectral notch indicating a substantial attenuation of sound around the crossover region. In this chapter, we show that a correct choice of the crossover frequency will influence the combined magnitude response *around the crossover region*.

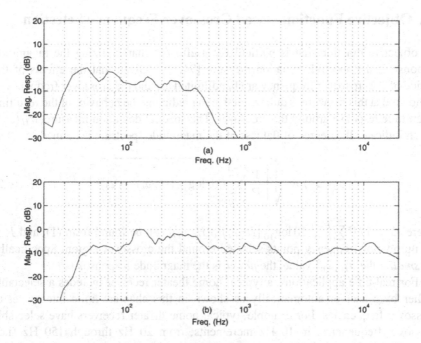

Fig. 6.5. (a) Magnitude response of the subwoofer measured in a reverberant room; (b) magnitude response of the satellite measured in the same room.

As an example, individual subwoofer and satellite (in this case a center channel) frequency responses (1/3rd octave smoothed), as measured in a room at a sampling frequency of 48 kHz with a reverberation time $T_{60} \approx .75$ sec, are shown in Figs. 6.5 (a) and (b), respectively. Clearly, the satellite is capable of playing audio below 100 Hz (up to about 40 Hz), whereas the subwoofer is most efficient and generally used for audio playback at frequencies less than 200 Hz. For example, as shown in Fig. 6.6, the resulting magnitude response, according to (6.1), obtained by summing the impulse responses, has a severe spectral notch for a crossover frequency ω_c corresponding to 60 Hz. This has been verified through real measurements where the subwoofer and the satellite channels were excited with a broadband stimuli (e.g., log-chirp signal) and subsequently deconvolving the net response from the measured signal.

Although room equalization has been widely used to solve problems in the magnitude response, the equalization filters do not necessarily solve the problems around the crossover frequency. In fact, many of these filters are minimum-phase and as such may do little to influence the result around the crossover. As shown in this chapter, automatic selection of a proper crossover frequency through an objective function allows the magnitude response to be flattened in the crossover region. All-pass-based optimization can overcome any additional limitations.

6.2 Objective Function-Based Crossover Frequency Selection

An objective function that is particularly useful for characterizing the magnitude response is the spectral deviation measure [94, 95]. Given that the effects of the choice of the crossover frequency are bandlimited around the crossover frequency, it is shown that this measure is quite effective in predicting the behavior of the resulting magnitude response around the crossover. The spectral deviation measure, $\sigma_H(\omega_c)$, which indicates the degree of flatness of the magnitude spectrum is defined as

$$\sigma_H(\omega_c) = \sqrt{\left[\frac{1}{P}\sum_{i=0}^{P-1}(10\log_{10}|H(\omega_i)| - \Delta)^2\right]} \qquad (6.2)$$

where $\Delta = 1/P\sum_{i=0}^{P-1}10\log_{10}|H(\omega_i)|$, $|H(\omega_i)|$ can be found from (1), and P is the number of frequency points selected around the crossover region. Specifically, the *smaller* the $\sigma_H(\omega_c)$ value, the *flatter* is the magnitude response.

For real-time applications, a typical home theater receiver includes a selectable (either by a user or automatically as shown in this chapter) finite integer set of crossover frequencies. For example, typical home theater receivers have selectable crossover frequencies, in 10 Hz increments, from 20 Hz through 150 Hz (i.e., $\Omega = [20\text{ Hz}, 30\text{ Hz}, 40\text{ Hz}, \ldots, 150\text{ Hz}]$). Thus, although a near-optimal solution ω_c^* can be found through a gradient descent optimization process by minimizing the spectral deviation measure with respect to ω (viz., $\partial\sigma_H(\omega_c)/\partial\omega_c|_{\omega_c=\omega_c^*}$), this is unnecessarily complicated. Clearly, the choice of the crossover frequency is limited to

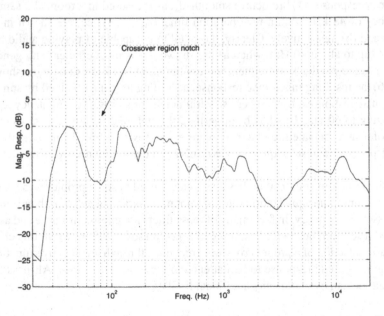

Fig. 6.6. Magnitude of the net response obtained from using a crossover frequency of 60 Hz.

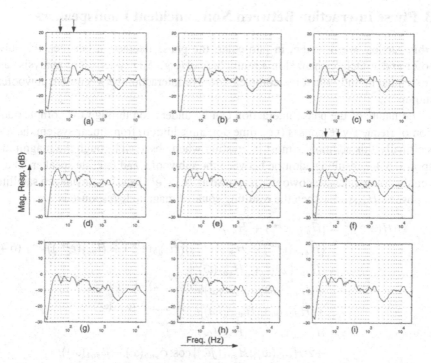

Fig. 6.7. Plots of the resulting magnitude response for crossover frequencies: (a) 50 Hz, (b) 60 Hz, (c) 70 Hz, (d) 80 Hz, (e) 90 Hz, (f) 100 Hz, (g) 110 Hz, (h) 120 Hz, (i) 130 Hz.

this finite set of integers (viz., as given in Ω), hence a simpler but yet effective means to select a proper choice of the crossover frequency, is to characterize the effect of the choice of each of the selectable integer crossover frequencies on the magnitude response in the crossover region.

Figure 6.7 shows the resulting magnitude responses, as obtained via (6.1), for different integer choices of the crossover frequencies from 50 Hz through 130 Hz. The corresponding spectral deviation values, as a function of the crossover frequency, for the crossover region around the crossover frequencies are shown in Fig. 6.8. Clearly, comparing Fig. 6.8 results with the plots in Figs. 6.7, it can be seen that the spectral deviation measure is an excellent measure for accurately modeling the performance in the crossover region for a given choice of crossover frequency. The best crossover frequency is then that which minimizes the spectral deviation measure, in the crossover region, over the integer set of crossover frequencies. Specifically,

$$\omega_c^* = \min_{\omega_c \in \Omega} \sigma_H(\omega_c) \qquad (6.3)$$

In this example 120 Hz provided the best choice for the crossover frequency as it gave the smallest $\sigma_H(\omega_c)$.

6.3 Phase Interaction Between Noncoincident Loudspeakers

In this section, we describe, in particular, the phase interaction between the subwoofer and a satellite channel in a multichannel (e.g., 5.1) system. The analysis can be extended to understand the complex additive interaction between nonsubwoofer channels.

The nature of the phase interaction can be understood through the complex addition of frequency responses (i.e., time domain addition) from linear system theory. Specifically, this addition is most interesting when observed through the magnitude response of the resulting addition between the subwoofer and satellite speaker. Thus, given the bass managed subwoofer response as $\tilde{H}_{sub}(e^{j\omega})$ and bass managed satellite response as $\tilde{H}_{sat}(e^{j\omega})$, then the resulting squared magnitude response is

$$
\begin{aligned}
|H(e^{j\omega})|^2 &= |\tilde{H}_{sub}(e^{j\omega}) + \tilde{H}_{sat}(e^{j\omega})|^2 \\
&= (\tilde{H}_{sub}(e^{j\omega}) + \tilde{H}_{sat}(e^{j\omega}))(\tilde{H}_{sub}(e^{j\omega}) + \tilde{H}_{sat}(e^{j\omega}))^{\dagger} \quad (6.4)\\
&= |\tilde{H}_{sub}(\omega)|^2 + |\tilde{H}_{sat}(\omega)|^2 \\
&\quad + |\tilde{H}_{sub}(\omega)||\tilde{H}_{sat}(j\omega)|e^{j(\phi_{sub}(\omega) - \phi_{sat}(\omega))} \\
&\quad + |\tilde{H}_{sub}(\omega)||\tilde{H}_{sat}(j\omega)|e^{-j(\phi_{sub}(\omega) - \phi_{sat}(\omega))} \\
&= |\tilde{H}_{sub}(\omega)|^2 + |\tilde{H}_{sat}(\omega)|^2 \\
&\quad + 2|\tilde{H}_{sub}(\omega)||\tilde{H}_{sat}(j\omega)| \cos(\phi_{sub}(\omega) - \phi_{sat}(\omega))
\end{aligned}
$$

where $\tilde{H}_{sat}(e^{j\omega})$ and $\tilde{H}_{sub}(e^{j\omega})$ are bass managed satellite and subwoofer channel room responses measured at a listening position in a room, and $A^{\dagger}(e^{j\omega})$ is the complex conjugate of $A(e^{j\omega})$. The phase responses of the subwoofer and the satellite are

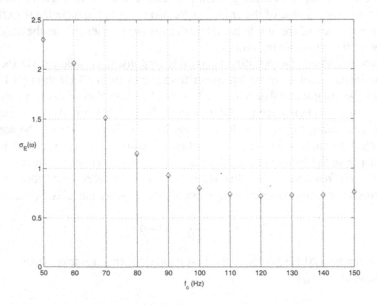

Fig. 6.8. Spectral deviation versus crossover frequency.

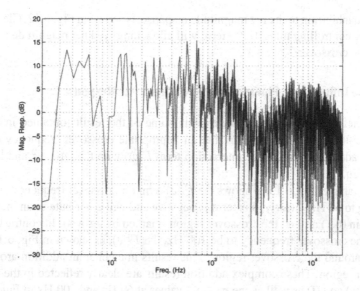

Fig. 6.9. Combined subwoofer satellite response at a particular listening position in a reverberant room.

given by $\phi_{sub}(\omega)$ and $\phi_{sat}(\omega)$, respectively. Furthermore, $\tilde{H}_{sat}(e^{j\omega})$ and $\tilde{H}_{sub}(e^{j\omega})$ may be expressed as

$$\tilde{H}_{sat}(e^{j\omega}) = BM_{sat}(e^{j\omega})H_{sat}(e^{j\omega})$$
$$\tilde{H}_{sub}(e^{j\omega}) = BM_{sub}(e^{j\omega})H_{sub}(e^{j\omega}) \tag{6.5}$$

where $BM_{sat}(e^{j\omega})$ and $BM_{sub}(e^{j\omega})$ are the the bass management IIR filters, whereas $H_{sat}(e^{j\omega})$ and $H_{sub}(e^{j\omega})$ are the full-range satellite and subwoofer responses, respectively.

The influence of phase on the net magnitude response is via the additive term $\Lambda(e^{j\omega}) = 2|\tilde{H}_{sub}(e^{j\omega})||\tilde{H}_{sat}(e^{j\omega})|\cos(\phi_{sub}(\omega) - \phi_{sat}(\omega))$. This term influences the combined magnitude response, generally, in a detrimental manner when it adds incoherently to the magnitude response sum of the satellite and the subwoofer. Specifically, when $\phi_{sub}(\omega) = \phi_{sat}(\omega) + k\pi$, $k = 1, 3, \ldots$, the resulting magnitude response is actually the difference between the magnitude responses of the subwoofer and the satellite thereby, possibly, introducing a spectral notch around the crossover frequency. For example, Fig. 6.9 shows an exemplary combined subwoofer center channel response in a room with reverberation time of about 0.75 seconds. Clearly, a large spectral notch is observed around the crossover, and one of the reasons for the introduction of this notch is the additive term $\Lambda(e^{j\omega})$ which adds incoherently to the magnitude response sum. Figure 6.10 is a third octave smoothed magnitude response corresponding to Fig. 6.9, whereas Fig. 6.11 shows the effect of the $\Lambda(e^{j\omega})$ term clearly exhibiting an inhibitory effect around the crossover region due to the phase interaction between the subwoofer and the satellite speaker response at the listening position. The cosine of the phase difference (viz., $(\phi_{sub}(\omega) - \phi_{sat}(\omega))$) that causes the inhibition to the net magnitude response, is shown in Fig. 6.12. Clearly, intelligently controlling this $\Lambda(e^{j\omega})$ term will allow improved net magnitude response around the crossover.

6.3.1 The Influence of Phase on the Net Magnitude Response

In this brief digression, we briefly explain some of the results obtained in Section 7.2. Specifically, we demonstrate that an appropriate crossover frequency enables coherent addition of the phase interaction term $\Gamma(\omega)$ with the $|A(\omega)|^2$ and $|B(\omega)|^2$ terms in Eq. (6.1).

For example, Fig. 6.13(b) shows the $\Gamma(\omega)$ term for crossover frequency ω_c corresponding to 60 Hz. Clearly, this term is negative and will contribute to an incoherent addition in (6.1) around the crossover region (marked by arrows). In contrast, by selecting the crossover frequency to be 100 Hz, the $\Gamma(\omega)$, as shown in Fig. 6.13(a), is positive around the crossover region. This results in a coherent addition around the crossover region. These complex addition results are clearly reflected in the plots of Figs. 6.7(a) and (f) as well as the $\sigma_H(\omega_c)$ values at 60 Hz and 100 Hz in Fig. 6.8.

6.4 Phase Equalization with All-Pass Filters

6.4.1 Second-Order All-Pass Networks

A second-order all-pass filter, $\mathcal{A}(z)$, can be expressed as

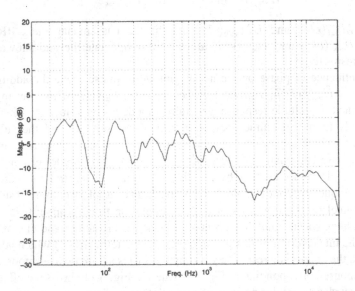

Fig. 6.10. The 1/3 octave smoothed combined magnitude response of Fig. 6.9.

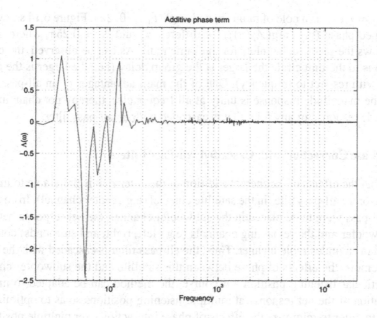

Fig. 6.11. The influence of $\Lambda(e^{j\omega})$ on the combined magnitude response.

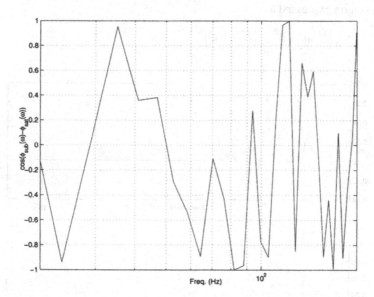

Fig. 6.12. Plot of the cosine of the phase difference that contributes to the incoherent addition around the crossover.

$$\mathcal{A}(z) = \left. \frac{z^{-1} - z_i^\dagger}{1 - z_i z^{-1}} \frac{z^{-1} - z_i}{1 - z_i^\dagger z^{-1}} \right|_{z=e^{j\omega}} \tag{6.6}$$

where $z_i = r_i e^{j\theta_i}$ is a pole of radius r_i and angle $\theta_i \in [0, 2\pi)$. Figure 6.14 shows the unwrapped phase (viz., $\arg(\mathcal{A}_p(z))$) for different r_i and $\theta_i = 0.25\pi$, whereas Fig. 6.15 shows the group delay plots for the same radii. As can be observed, the closer the pole is to the unit circle the larger is the group delay (i.e., the larger is the phase change with respect to frequency). One of the main advantages of an all-pass filter is that the magnitude response is unity at all frequencies, thereby not changing the magnitude response of any filter that is cascaded with an all-pass filter.

6.4.2 Phase Correction with Cascaded All-Pass Filters

To combat the effects of incoherent addition of the Λ term, it is preferable to include this first-order all-pass filter in the satellite channel (e.g., center channel). In contrast, if the all-pass were to be placed in the subwoofer channel, the net response between the subwoofer and the remaining channels (e.g., left, right, and surrounds) could be affected in an undesirable manner. Thus, the all-pass filter is cascaded with the satellite to remove the effects of phase between this satellite and the subwoofer channel at a particular listening position. Of course, the method can be adapted to include information of the net response at multiple listening positions so as to optimize the Λ term in order to minimize the effects of phase interaction over multiple positions.

Now, the net response with an M-cascade all-pass filter, $\mathcal{A}_M(e^{j\omega})$, in the satellite channel, can be expressed as

$$|H(e^{j\omega})|^2 = |\tilde{H}_{sub}(\omega)|^2 + |\tilde{H}_{sat}(\omega)|^2$$

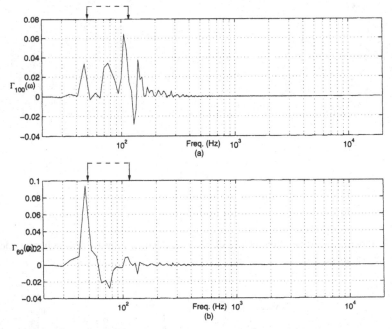

Fig. 6.13. $\Gamma(\omega)$ term from (6.1) for crossover frequency (a) 100 Hz, (b) 60 Hz.

$$+2|\tilde{H}_{sub}(\omega)||\tilde{H}_{sat}(j\omega)|\cos(\phi_{sub}(\omega)-\phi_{sat}(\omega)-\phi_{\mathcal{A}_M}(\omega)) \quad (6.7)$$

where

$$\mathcal{A}_M(e^{j\omega}) = \prod_{k=1}^{M} \frac{e^{-j\omega}-r_k e^{-j\theta_k}}{1-r_k e^{j\theta_k}e^{-j\omega}}\frac{e^{-j\omega}-r_k e^{j\theta_k}}{1-r_k e^{-j\theta_k}e^{-j\omega}}$$

$$\phi_{\mathcal{A}_M}(\omega) = \sum_{k=1}^{M} \phi_{\mathcal{A}_M}^{(k)}(\omega) \quad (6.8)$$

$$\phi_{\mathcal{A}_M}^{(i)}(\omega) = -2\omega - 2\tan^{-1}\left(\frac{r_i\sin(\omega-\theta_i)}{1-r_i\cos(\omega-\theta_i)}\right)$$

$$-2\tan^{-1}\left(\frac{r_i\sin(\omega+\theta_i)}{1-r_i\cos(\omega+\theta_i)}\right)$$

and $\Lambda_F(e^{j\omega}) = 2|\tilde{H}_{sub}(e^{j\omega})||\tilde{H}_{sat}(e^{j\omega})|\cos(\phi_{sub}(\omega)-\phi_{sat}(\omega)-\phi_{\mathcal{A}_M}(\omega))$. Thus, to minimize the inhibitory effect of Λ term (or in effect cause it to coherently add to $|\tilde{H}_{sub}(\omega)|^2 + |\tilde{H}_{sat}(\omega)|^2$), in the example above, one can define an average square error function (or objective function) for minimization as

$$J(n) = \frac{1}{N}\sum_{l=1}^{N} W(\omega_l)(\phi_{sub}(\omega)-\phi_{sat}(\omega)-\phi_{\mathcal{A}_M}(\omega))^2 \quad (6.9)$$

where $W(\omega_l)$ is a frequency-dependent weighting function.

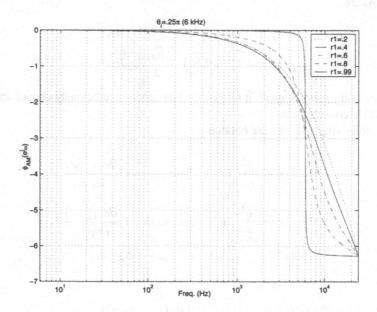

Fig. 6.14. Plot of the unwrapped phase of a second-order all-pass filter for different values of the pole magnitude for $\theta = 0.25\pi$.

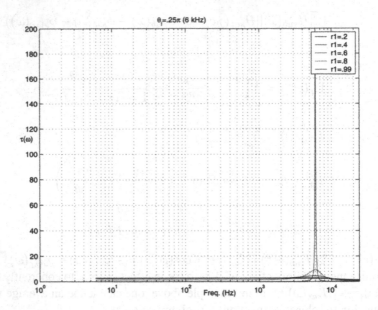

Fig. 6.15. Plot of the group delay of a second-order all-pass filter for different values of the pole magnitude for $\theta = 0.25\pi$.

The terms r_i and $\theta_i (i = 1, 2, \ldots, M)$ can be determined adaptively by minimizing the objective function with respect to these unknown parameters. The update equations are

$$r_i(n+1) = r_i(n) - \frac{\mu_r}{2}\nabla_{r_i}J(n)$$

$$\theta_i(n+1) = \theta_i(n) - \frac{\mu_\theta}{2}\nabla_{\theta_i}J(n) \tag{6.10}$$

where μ_r and μ_θ are adaptation rate control parameters judiciously chosen to guarantee stable convergence.

The following relations are obtained.

$$\nabla_{r_i}J(n) = \sum_{l=1}^{N} W(\omega_l)E(\phi(\omega))(-1)\frac{\partial\phi_{A_M}(\omega)}{\partial r_i(n)}$$

$$\nabla_{\theta_i}J(n) = \sum_{l=1}^{N} W(\omega_l)E(\phi(\omega))(-1)\frac{\partial\phi_{A_M}(\omega)}{\partial\theta_i(n)}$$

$$E(\phi(\omega)) = \phi_{sub}(\omega) - \phi_{sat}(\omega) - \phi_{A_M}(\omega) \tag{6.11}$$

where

$$\frac{\partial\phi_{A_M}(\omega)}{\partial r_i(n)} = -\frac{2\sin(\omega_l - \theta_i(n))}{r_i^2(n) - 2r_i(n)\cos(\omega_l - \theta_i(n)) + 1}$$

$$-\frac{2\sin(\omega_l + \theta_i(n))}{r_i^2(n) - 2r_i(n)\cos(\omega_l + \theta_i(n)) + 1} \tag{6.12}$$

and

$$\frac{\partial \phi_{A_M}(\omega)}{\partial \theta_i(n)} = -\frac{2r_i(n)(r_i(n) - \cos(\omega_l - \theta_i(n)))}{r_i^2(n) - 2r_i(n)\cos(\omega_l - \theta_i(n)) + 1}$$
$$-\frac{2r_i(n)(r_i(n) - \cos(\omega_l + \theta_i(n)))}{r_i^2(n) - 2r_i(n)\cos(\omega_l + \theta_i(n)) + 1} \tag{6.13}$$

During the update process, care was taken to ensure that $|r_i(n)| < 1$ to guarantee stability. This was done by including a condition where the r_i element that exceeded unity would be randomized. Clearly, this could increase the convergence time, and hence in the future other methods may be investigated to minimize the number of iterations for determining the solution.

6.4.3 Results

For the combined subwoofer center channel response shown in Fig. 6.9, the r_i and θ_i with $M = 9$ were adapted to get reasonable minimization of $J(n)$. Furthermore, the frequency dependent weighting function, $W(\omega_l)$, for the above example was chosen as unity for frequencies between 60 Hz and 125 Hz. The reason for this choice of weighting terms can be readily seen from the domain of the $\Lambda(e^{j\omega})$ term of Fig. 6.11 and/or the domain of the "suckout" in Fig. 6.10.

The original phase difference function $(\phi_{sub}(\omega) - \phi_{sat}(\omega))^2$ is plotted in Fig. 6.16 and the cosine term, $\cos(\phi_{sub}(\omega) - \phi_{sat}(\omega))$ that adds incoherently is shown in Fig. 6.12. Clearly, minimizing the phase difference (using the all-pass cascade in the satellite channel) around the crossover region will minimize the spectral notch. The resulting all-pass filtered phase difference function, $(\phi_{sub}(\omega) - \phi_{sat}(\omega) - \phi_{A_M}(\omega))^2$, from the adaptation of $r_i(n)$ and $\theta_i(n)$ is shown in Fig. 6.17 thereby demonstrating the minimization of the phase difference around the crossover. The resulting all-pass filtered term, $\Lambda_F(\omega)$, is shown in Fig. 6.18. Comparing Figs. 6.11 and 6.18, it can be seen that the inhibition turns to an excitation to the net magnitude response around the crossover region. Finally, Fig. 6.19 shows the resulting combined magnitude response with the cascade all-pass filter in the satellite channel, and Fig. 6.20 shows the third octave smoothed version of Fig. 6.19. A superimposed plot, comprising Fig. 6.20 and the original combined response of Fig. 6.10 is depicted in Fig. 6.21. Clearly, an improvement of about 7 dB around the crossover can be seen.

6.5 Objective Function-Based Bass Management Filter Parameter Optimization

A typical equalization filter design process involves (i) measuring the loudspeaker and room responses for each of the satellites and the subwoofer, (ii) storing these responses, (iii) designing an equalization filter for each channel loudspeaker (viz., warping and LPC based as shown in Fig. 6.22), and (iii) applying individual bass management filters to each of the equalization filters in a multichannel audio system.

Quantitatively, as an example, the net subwoofer and satellite response at a listening position, as shown by Fig. 6.23, can be expressed as

$$|H(e^{j\omega})|^2 = |H_{sub}(e^{j\omega})\tilde{H}_{sub}^{-1}(e^{j\omega}) + H_{sat}(e^{j\omega})\tilde{H}_{sat}^{-1}(e^{j\omega})|^2 \qquad (6.14)$$
$$= |H_{sub}(e^{j\omega})\tilde{H}_{sub}^{-1}(e^{j\omega})|^2 + |H_{sat}(e^{j\omega})\tilde{H}_{sat}^{-1}(e^{j\omega})|^2$$
$$+ 2|H_{sub}(e^{j\omega})\tilde{H}_{sub}^{-1}(e^{j\omega})||H_{sat}(e^{j\omega})\tilde{H}_{sat}^{-1}(e^{j\omega})|$$
$$* \cos(\phi_{sub}(\omega) + \tilde{\phi}_{sub}(\omega) - \phi_{sat}(\omega) - \tilde{\phi}_{sat}(\omega))$$

where $\tilde{H}_{sat}^{-1}(e^{j\omega})$ and $\tilde{H}_{sub}^{-1}(e^{j\omega})$ are bass managed equalization filters for the satellite and subwoofer channel responses measured at a listening position in a room. The phase responses of the subwoofer and the satellite bass managed filters are given by $\tilde{\phi}_{sub}(\omega)$ and $\tilde{\phi}_{sat}(\omega)$, respectively. Specifically, $\tilde{H}_{sat}^{-1}(e^{j\omega})$, $\tilde{H}_{sub}^{-1}(e^{j\omega})$, $\tilde{\phi}_{sub}(\omega)$, and $\tilde{\phi}_{sat}(\omega)$ may be expressed as

$$\tilde{H}_{sat}^{-1}(e^{j\omega}) = H_{\omega_c,N}^{hp}(e^{j\omega})\hat{H}_{sat}^{-1}(e^{j\omega})$$
$$\tilde{H}_{sub}^{-1}(e^{j\omega}) = H_{\omega_c,M}^{lp}(e^{j\omega})\hat{H}_{sub}^{-1}(e^{j\omega}) \qquad (6.15)$$
$$\tilde{\phi}_{sat}(\omega) = -\hat{\phi}_{sat}(\omega) + \phi_{\omega_c,N}^{hp}(\omega)$$
$$\tilde{\phi}_{sub}(\omega) = -\hat{\phi}_{sub}(\omega) + \phi_{\omega_c,M}^{lp}(\omega)$$

where the "hat" above the frequency and phase responses, of the subwoofer and satellite equalization filter, represents an approximation due to the lower order spec-

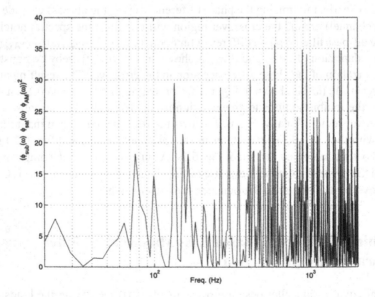

Fig. 6.16. Plot of the phase difference, $(\phi_{sub}(\omega) - \phi_{sat}(\omega))^2$, as a function of frequency.

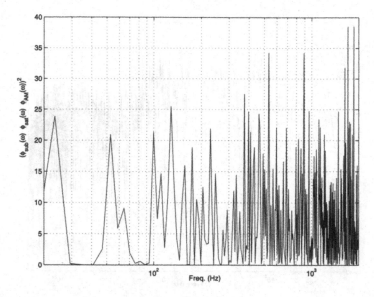

Fig. 6.17. Plot of the all-pass filtered phase difference, $(\phi_{sub}(\omega) - \phi_{sat}(\omega) - \phi_{A_M}(\omega))^2$, as a function of frequency. Observe the reduction around the crossover (≈ 80 Hz).

tral modeling via LPC. As is evident from Eqs. (6.14) and (6.15), the crossover region

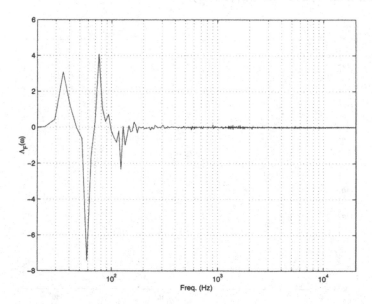

Fig. 6.18. The influence of the all-pass filtered function $\Lambda_F(e^{j\omega})$ on the combined magnitude response.

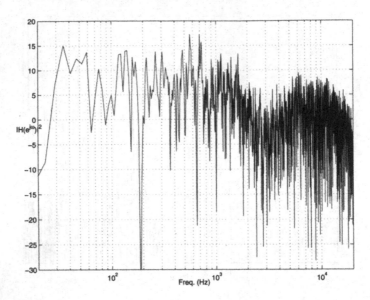

Fig. 6.19. The combined magnitude response of the subwoofer and the satellite with a cascade of all-pass filters.

response of $|H(e^{j\omega})|^2$ can be further optimized through a proper choice of the bass management filter parameters (ω_c, N, M).

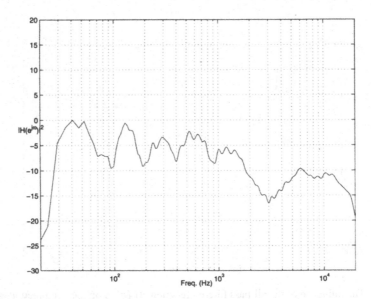

Fig. 6.20. The 1/3 octave smoothed combined magnitude response of Fig. 6.19.

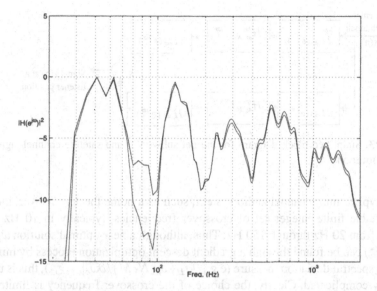

Fig. 6.21. A superimposed plot of the original subwoofer and satellite combined magnitude response and the all-pass filter-based combined magnitude response demonstrating at least abut 7 dB improvement around the crossover region ($\approx$80 Hz).

An objective function that is particularly useful for characterizing the magnitude response is the spectral deviation measure [94, 95]. Given that the effects of the choice of the bass management parameters (viz., (ω_c, N, M)) are bandlimited around the crossover frequency, it is shown that that this measure is quite effective in predicting the behavior of the resulting magnitude response around the crossover. The spectral deviation measure, $\sigma_H(\omega_c, N, M)$, which indicates the degree of flatness of the magnitude spectrum is defined as

$$\sigma_H(\omega_c, N, M) = \sqrt{\left[\frac{1}{P}\sum_{i=0}^{P-1}(10\log_{10}|H(e^{j\omega_i})| - \Delta)^2\right]} \quad (6.16)$$

where $\Delta = 1/P\sum_{i=0}^{P-1}10\log_{10}|H(e^{j\omega_i})|$, $|H(e^{j\omega_i})|$ can be found from Eq. (6.14), and P is the number of frequency points selected around the crossover region. Specifically, the *smaller* the $\sigma_H(\omega_c)$ value, the *flatter* is the magnitude response around the crossover region.

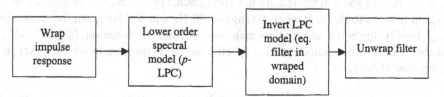

Fig. 6.22. Warping based equalization.

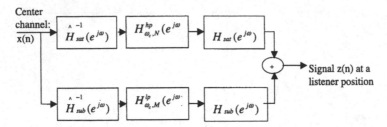

Fig. 6.23. Simplified block diagram for the net subwoofer and satellite channel signal at a listener location.

A typical multichannel audio system, such as a home theater receiver, includes a selectable finite integer set of crossover frequencies, typically in 10 Hz increments, from 20 Hz through 150 Hz. Thus, although a near-optimal solution ω_c^* (and N^*, M^*) can be found through a gradient descent optimization process by minimizing the spectral deviation measure (e.g., $\partial\sigma_H(\omega_c, N, M)/\partial\omega_c|_{\omega_c=\omega_c^*}$), this is unnecessarily complicated. Clearly, the choice of the crossover frequency is limited to a finite set of integers. Thus, a simple but effective means to select a proper choice of the crossover frequency, is to characterize the effect of the choice of the selected crossover frequency on $\sigma_H(\omega_c, N, M)$ in the crossover region. Similarly, the choice of N, M can essentially be limited to a finite set of integers. Thus, the effect of varying (ω_c, N, M), *jointly*, can be immediately observed on $\sigma_H(\omega_c, N, M)$.

6.5.1 Results

The full-range subwoofer and satellite responses of Fig. 6.24 were used for obtaining the corresponding equalization filters with the warping and LPC modeling method of Fig. 6.22.

The integer bass management parameters to be applied to the equalization filters, $\hat{H}_{sat}^{-1}(e^{j\omega})$ and $\hat{H}_{sub}^{-1}(e^{j\omega})$, were selected from the following intervals: $\omega_c \in \{50, 150\}$, $N \in \{1, 5\}$, $M \in \{1, 4\}$. Subsequently, for a particular combination of (ω_c, N, M), the bass managed equalization filters, $\tilde{H}_{sat}^{-1}(e^{j\omega})$ and $\tilde{H}_{sub}^{-1}(e^{j\omega})$, were applied to $H_{sub}(e^{j\omega})$ and $H_{sat}(e^{j\omega})$ to yield the equalized responses $H_{sub}(e^{j\omega}) *$ $\tilde{H}_{sub}^{-1}(e^{j\omega})$ and $H_{sat}(e^{j\omega})\tilde{H}_{sub}^{-1}(e^{j\omega})$. The equalized subwoofer response was 1/3 octave smoothed and level matched with the 1/3 octave smoothed equalized satellite response. The resulting complex frequency response, $H(e^{j\omega})$, and the corresponding magnitude response were obtained. At this point, the spectral deviation measure $\sigma_H(\omega_c, N, M)$ was determined, for the given choice of (ω_c, N, M), in the crossover region (chosen to be the frequency range of 40 Hz and 200 Hz given the choice of ω_c). Finally, the best choice of bass management filter parameter set, (ω_c^*, N^*, M^*), is then that set which minimizes the spectral deviation measure in the crossover region. Specifically,

$$(\omega_c^*, N^*, M^*) = \min_{\omega_c, N, M} \sigma_H(\omega_c, N, M) \qquad (6.17)$$

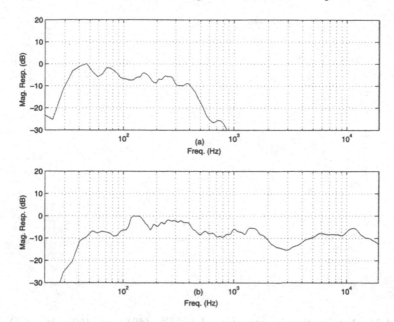

Fig. 6.24. (a) 1/3 octave smoothed magnitude response of the subwoofer-based LRTF measured in a reverberant room; (b) 1/3 octave smoothed magnitude response of the satellite-based LRTF measured in the same room.

The lowest spectral deviation measure, $\sigma_H^*(\omega_c, N, M)$, was obtained for (ω_c^*, N^*, M^*) corresponding to (60 Hz, 3, 4). Observing the natural full-range 18 dB/octave approximate decay rate (below 60 Hz) of the satellite, in Fig. 6.24(b), it is evident that this choice of $N = 3$ (i.e., 18 dB/octave roll-off to the satellite speaker equalized response) will not cause the satellite speaker to be distorted. If necessary, in the event that N is not sufficiently high, the next largest $\sigma_H(\omega_c, N', M)$ can always be selected such that $N' > N$. Of course, other signal-limiting mechanisms may be employed in conjunction with the proposed approach, and these are beyond the scope of this chapter. For this choice of the bass management filter parameters (i.e., (60 Hz, 3, 4)), the net magnitude response $|H(e^{j\omega})|^2$ (in dB) is shown in Fig. 6.25. Clearly, the variation in the crossover region (viz., 40 Hz through 200 Hz) is negligible, and this is reflected by the smallest value found for $\sigma_H^*(\omega_c, N, M) = 0.45$. Thus, the parameter set (60 Hz, 3, 4) forms the correct choice for the bass management filters for the room responses of Figs. 6.24.

Further examples, as provided in Figs. 6.26 to 6.28, show the net magnitude response $|H(e^{j\omega})|^2$ for different choices of (ω_c, N, M) that produce a larger $\sigma_H(\omega_c, N, M)$. As can be seen, these "nonoptimal" integer choices of the bass management filter parameters, as determined from the spectral deviation measure, cause significant variations in the magnitude response in the crossover region.

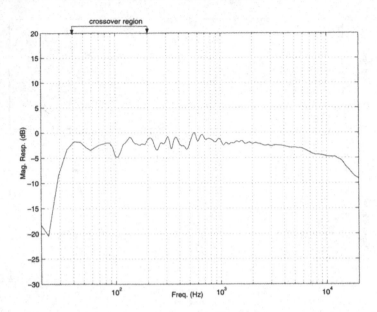

Fig. 6.25. Net magnitude response $|H(e^{j\omega})|^2$ (dB) for (60 Hz, 3, 4) with $\sigma_H^*(0.0025\pi, 3, 4) = 0.45$.

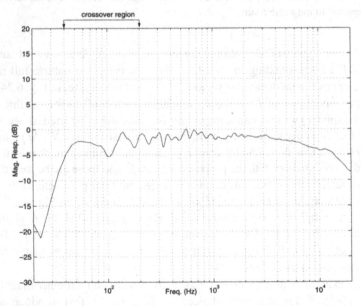

Fig. 6.26. Net magnitude response $|H(e^{j\omega})|^2$ (dB) for (50 Hz, 4, 4) with $\sigma_H(0.0021\pi, 4, 4) = 0.61$.

6.6 Multiposition Bass Management Filter Parameter Optimization

For multiposition bass management parameter optimization, an average spectral deviation measure can be expressed as

$$\sigma_H^{avg} = \frac{1}{L} \sum_{j=1}^{L} \sigma_{H_j}(\omega_c, N, M) \tag{6.18}$$

where L is the total number of positions equalized during the multiposition equalization step.

In a nutshell the bass management parameters can be optimized by the following steps, (i) perform multiple position equalization on raw responses (i.e., responses without any bass managed applied to them), (ii) apply the candidate bass management filters to the equalized subwoofer and satellite responses, parameterized by (ω_c, N, M) with $f_c \in \{40, 200\}$ Hz, $N \in \{2, 4\}$, $M \in \{3, 5\}$, (iii) perform subwoofer and satellite level setting using bandlimited noise and perceptual C-weighting, (iv) determine the average spectral deviation measure, σ_H^{avg}, after performing 1/3 octave smoothing for each of the net (i.e., combined subwoofer and satellite) responses in the range of 40 Hz and 250 Hz, and (v) select (ω_c, N, M) that minimizes σ_H^{avg}.

6.6.1 Results

As a first example the full-range subwoofer and satellite responses at $L = 4$ positions measured in a reverberant room with $T_{60} \approx 0.75s$, as shown in Fig. 6.29, were used for obtaining the corresponding equalization filters with the pattern recognition, warping, and LPC modeling method. The integer bass management parameters to be applied to the equalization filters, $\hat{H}_{sat}^{-1}(\omega)$ and $\hat{H}_{sub}^{-1}(\omega)$, and the general steps for

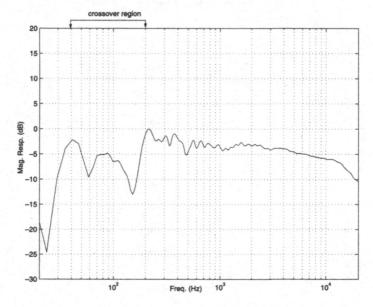

Fig. 6.27. Net magnitude response $|H(e^{j\omega})|^2$ (dB) for $(130\,\text{Hz}, 2, 1)$ with $\sigma_H(0.0054\pi, 2, 1) = 1.56$.

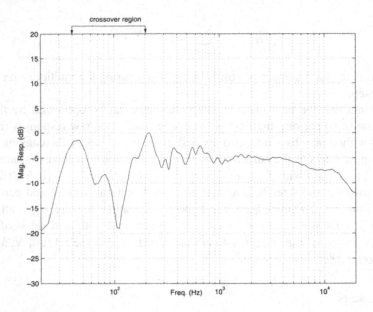

Fig. 6.28. Net magnitude response $|H(e^{j\omega})|^2$ (dB) for (90 Hz, 5, 3) with $\sigma_H(0.0037\pi, 5, 3) = 2.52$.

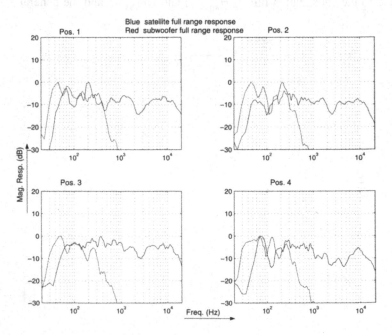

Fig. 6.29. An example of full-range subwoofer and satellite responses measured at four different positions in a room with $T_{60} \approx 0.75$ s.

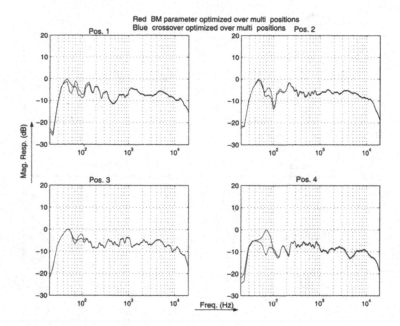

Fig. 6.30. The equalized and bass management filters parameter optimized responses where the lighter (thin) curve corresponds to all parameter optimization and the thick or darker curve corresponds to crossover frequency optimization over multiple positions.

determining the parameter set, (ω_c, N, M), that minimizes σ_H^{avg} are described in the preceding section. As a comparison, we also present the results of performing only crossover frequency optimization [96], but for multiple positions, using the average spectral deviation measure. The resulting equalized plots are shown in Fig. 6.30.

Comparing the results using the full parameter optimization (lighter curve) with the crossover frequency optimization over multiple positions, it can be seen that all-parameter flattens the magnitude response around the crossover region (viz., 40 Hz through 250 Hz). Specifically, for example, in position 2 a lower Q (i.e., broad) notch around the crossover region obtained through crossover frequency optimization is transformed to a high Q (i.e., narrow width) notch by all-parameter optimization. In addition, as shown via position 4, a very broad and high amplitude undesirable peak in the magnitude response, obtained from crossover frequency optimization, is reduced in amplitude and narrowed through all-parameter optimization (red curve). In fact, Toole and Olive [97] have demonstrated that, based on steady-state measurements, low-Q resonances producing broad peaks in the measurements are more easily heard than high-Q resonances producing narrow peaks of similar amplitude. The crossover frequency optimization resulted in a crossover at 90 Hz with the minimum of the average spectral deviation measure, σ_H^{avg}, being 0.98. The all-parameter optimization resulted in the parameter set (ω_c, N, M) corresponding to (80 Hz, 4, 5) and σ_H^{avg} for this parameter set being minimum at 0.89. Also, a comparison between

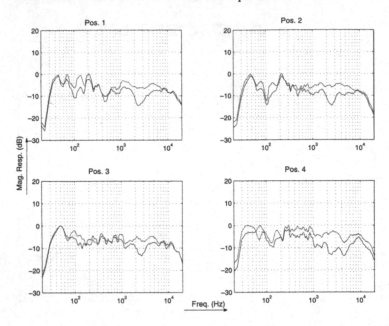

Fig. 6.31. The equalized and bass management filters parameter optimized responses where the thin curve corresponds to all-parameter optimization and the thick curve corresponds to unequalized responses with the standard bass management filters (80 Hz, 4, 5).

un-equalized responses with bass management set to the standard of (80 Hz, 4, 5) and the all-parameter optimized and equalized responses is shown in Fig. 6.31.

6.7 Spectral Deviation and Time Delay-Based Correction

The crossover region can be manipulated through the use of a simple delay in each loudspeaker (i.e., nonsubwoofer) channel. Specifically the output signal $y(n)$ can be expressed in terms of the input signal $x(n)$, the satellite bass management filter $bm_{sat}(n)$, the subwoofer bass management filter $bm_{sub}(n)$, the room responses $h_{sat}(n)$ and $h_{sub}(n)$, and the delay of n_d samples (viz., $\delta(n - n_d)$),

$$
\begin{aligned}
y(n) = &\, bm_{sat}(n) \otimes \delta(n - n_d) \otimes h_{sat}(n) \otimes x(n) \\
&+ bm_{sub}(n) \otimes h_{sub}(n) \otimes x(n)
\end{aligned}
\tag{6.19}
$$

The frequency domain representation for the resulting response leads to

$$
\begin{aligned}
H(e^{j\omega}) = &\, BM_{sat}(e^{j\omega}) e^{-j\omega n_d} H_{sat}(e^{j\omega}) \\
&+ BM_{sub}(e^{j\omega}) H_{sub}(e^{j\omega})
\end{aligned}
\tag{6.20}
$$

whereas the system representation is shown in Fig. 6.32.

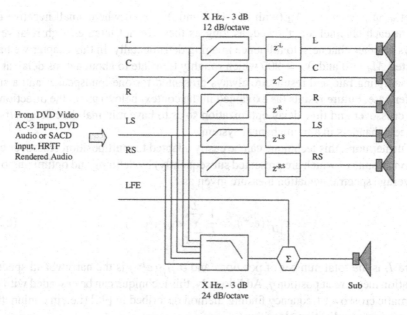

Fig. 6.32. System representation of time delay technique to correct crossover region response.

An objective function that is particularly useful for characterizing the magnitude response variations is the spectral deviation measure,

$$\sigma_{|H|}(e^{j\omega}) = \sqrt{\left[\frac{1}{D} \sum_{i=P_1}^{P_2} (10 \log_{10} |H(e^{j\omega_i})| - \Delta)^2 \right]} \qquad (6.21)$$

where $\Delta = 1/P \sum_{i=P_1}^{P_2} 10 \log_{10} |H(e^{j\omega_i})|$, $D = (P_2 - P_1 + 1)$ and $|H(e^{j\omega_i})|$ can be found from (6.20), and P is the number of frequency points selected around the crossover region. For the present simulations, because the bass management filters that were selected had a crossover around 80 Hz, P_1 and P_2 were selected to be bin numbers corresponding to 40 Hz and 200 Hz, respectively (viz., for an 8192 length response and a sampling rate of 48 kHz, the bin numbers corresponded to about 6 and 34 for 40 Hz and 200 Hz, respectively).

Thus, the process for selecting the best time delay n_d^* is: (i) set $n_d = 0$ (this may be relative to any delay that may be used for time-aligning speakers such that the relative delays between signals from various channels to a listening position is approximately zero), (ii) level match the subwoofer and satellite, (iii) determine (6.19), (iv) determine (6.21), (v) $n_d = n_d + 1$, (vi) perform (ii) to (v) until $n_d < N_d$, and (vii) select $n_d^* = \min \sigma_{|H|}(e^{j\omega})$.

Care should be taken to ensure that (i) the delay n_d is not large enough to cause a perceptible delay between audio and video frames, and (ii) the relative delays between channels should not be large enough to cause any imaging problems. Furthermore, if n_d is set relative to time-aligned delays, then the termination condition can be set as $M_d < n_d < N_d$ (with $M_d < 0$ and $N_d > 0$) where small negative delays in each channel are allowed, as long as they are not large enough relative to delays in other channels, to influence imaging detrimentally. In this chapter we have selected $M_d = 0$ and $N_d = 200$ which roughly translate to about a 4 ms delay at 48 kHz sampling rate and results are always presented for one loudspeaker and a subwoofer case. Future results (as explained in the context below) are in the direction of joint crossover and time delay optimization so as to have minimal time delay offsets between channels in a multichannel system.

Furthermore, this technique can be easily adapted to multiposition crossover correction (results of which are presented subsequently) by defining and optimizing over an average spectral deviation measure given as

$$\sigma_{|H|}^{avg}(e^{j\omega}) = \frac{1}{L} \sum_{j=1}^{L} \sigma_{|H_j|}(e^{j\omega}) \tag{6.22}$$

where L is the total number of positions and $\sigma_{|H_j|}(e^{j\omega})$ is the narrowband spectral deviation measure at position j. Additionally, this technique can be cascaded with the automatic crossover frequency finding method described in [98] (i.e., in conjunction with a room equalization algorithm).

6.7.1 Results for Spectral Deviation and Time Delay-Based Crossover Correction

Figure 6.33 shows the full-range satellite and subwoofer response at a listening position, whereas Fig. 6.34 compares the bass managed response (dash-dot line), with crossover at 60 Hz, with the spectral deviation-based time delay corrected crossover region response. The optimal time delay found was $n_d^* = 142$ samples at 48 kHz.

Figure 6.35 compares the correction being done in the crossover region using the automatic time delay and spectral deviation-based technique but for a crossover of 70 Hz for the same speaker set and same position as that of Fig. 6.33. The optimal time delay was 88 samples. A minimal 10 Hz difference in the crossover required an additional 54 sample delay to correct the crossover region response. One possibility is that this is because the suckout for the 70 Hz case was less deep than the 60 Hz case and hence needed a smaller time delay for effective crossover correction. This is further validated by selecting the crossover to be 90 Hz and observing the time delay correction required. Figure 6.36 shows that a small amount of crossover region response correction is achieved, for crossover of 90 Hz, given the optimal time delay $n_d^* = 40$ samples. This time delay is a further reduction of 48 samples over the 70 Hz case as the crossover region response is further optimized when the crossover frequency is selected at 90 Hz. From the three crossover frequencies for the center-sub case, 90 Hz is the best crossover frequency as it gives the least amount of suckout in the crossover region and hence requires a very small time delay of just 40 samples.

Accordingly, it can be inferred that the time delay offsets between channels in a multichannel setup can be kept at a minimum, but still provide crossover region correction, by either of the following techniques, (i) first performing a crossover frequency search by the crossover finding method to improve the crossover region response for each channel loudspeaker and subwoofer response and then applying a relatively smaller time delay correction to each satellite channel to further improve the crossover response, or (ii) performing a multidimensional search for the best choice of time delay and the crossover, simultaneously, using the spectral deviation measure, so as to keep the time delay offsets between channels at a minimum.

Figure 6.37 shows the full-range subwoofer and left surround responses at a listening position, whereas Fig. 6.38 compares the bass managed response (dash-dot line), with crossover at 120 Hz, with the spectral deviation-based time delay corrected crossover region response. The optimal time delay n_d^* was 73 samples at 48 kHz.

6.8 Summary

In this chapter we presented results that show the effect of a proper choice of crossover frequency for improving low-frequency performance. Additional parameter optimization of the bass management filters is shown to yield improved performance. Comparison between the results from crossover and all-parameter optimization, of the bass management filters, for multiposition equalization is presented. As was shown, cascading an all-pass filter can provide further improvements to the equalization result in the crossover region. Alternatively, time delay adjustments can be made in each loudspeaker channel to correct the crossover region response.

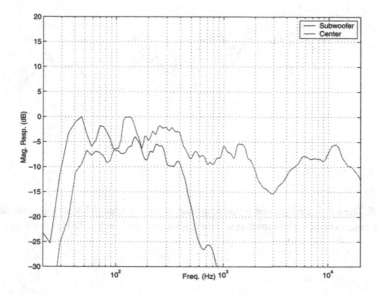

Fig. 6.33. The individual full-range subwoofer and a center channel magnitude response measured at a listening position in a reverberant room.

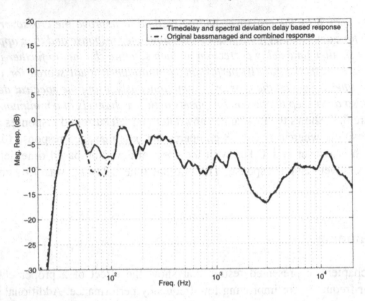

Fig. 6.34. The bass managed combined response as well as time delay and spectral deviation measure-based corrected crossover response (crossover frequency = 60 Hz).

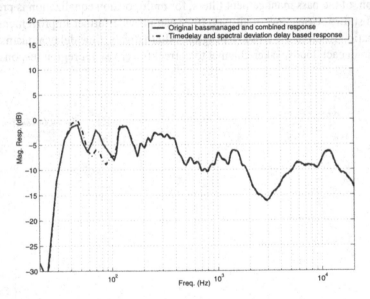

Fig. 6.35. The bass managed combined response as well as time delay and spectral deviation measure-based corrected crossover response (crossover frequency = 70 Hz).

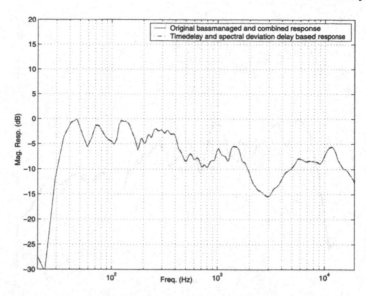

Fig. 6.36. The bass managed combined response as well as time delay and spectral deviation measure-based corrected crossover response for the left surround (crossover frequency = 90 Hz).

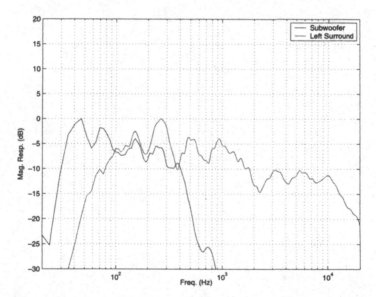

Fig. 6.37. The individual full-range subwoofer and a left surround channel magnitude response measured at a listening position in a reverberant room.

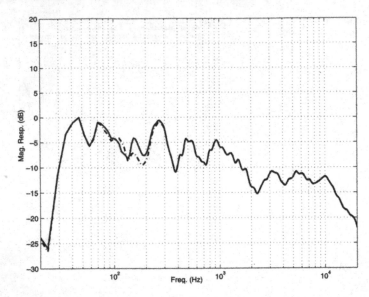

Fig. 6.38. The bass managed combined response as well as time delay and spectral deviation measure-based corrected crossover response for the left surround (crossover frequency = 120 Hz).

7

Robustness of Equalization to Displacement Effects: Part I

Traditionally, multiple listener room equalization is performed to improve sound quality at all listeners, during audio playback, in a multiple listener environment (e.g., movie theaters, automobiles, etc.). A typical way of doing multiple listener equalization is through spatial averaging, where the room responses are averaged spatially between positions and an inverse equalization filter is found from the spatially averaged result. However, the equalization performance will be affected if there is a mismatch between the position of the microphones (which are used for measuring the room responses for designing the equalization filter) and the actual center of listener head position (during playback). In this chapter, we present results of the effects of microphone and listener mismatch on spatial average equalization performance for frequencies above the Schroeder frequency. The results indicate that, for the analyzed rectangular listener configuration, the region of effective equalization depends on (i) distance of a listener from the source, (ii) amount of mismatch between the responses, and (iii) the frequency of the audio signal. We also present some convergence analysis to interpret the results.

7.1 Introduction

A typical room is an acoustic enclosure that can be modeled as a linear system whose behavior at a particular listening position is characterized by an impulse response. The impulse response yields a complete description of the changes a sound signal undergoes when it travels from a source to a receiver (microphone/listener). The signal at the receiver consists of direct path components, discrete reflections that arrive a few milliseconds after the direct sound, as well as a reverberant field component. In addition, it is well established that room responses change with source and receiver locations in a room [11, 63].

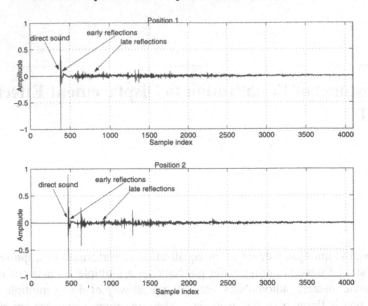

Fig. 7.1. Examples of room acoustical responses, having the direct and reverberant components, measured at two positions a few feet in a room.

Specifically, the time of arrival of the direct and multipath reflections and the energy of the reverberant component will vary from position to position. In other words, a room response at position i, $p_{f,i}$, can be expressed as $p_{f,i} = p_{f,d,i} + p_{f,rev,i}$; whereas the room response at position j, $p_{f,j}$, can be expressed as $p_{f,j} = p_{f,d,j} + p_{f,rev,j}$ where $p_{f,d,j}$ is the frequency response for the direct path component, and $p_{f,rev,j}$ is the response for the multipath component. An example of time domain responses at two positions, displaced a few feet apart from each other, in a room with reverberation time of about 0.25 seconds, is shown in Fig. 7.1 along with the direct component, early reflections, and late reverberant components. Figure 7.2 shows the corresponding frequency response from 20 Hz to 20 kHz.

One of the goals in equalization is to minimize the spectral deviations (viz., correcting the peaks and dips) found in the magnitude response through an equalization filter. This correction of the room response significantly improves the quality of sound played back through a loudspeaker system. In essence, the resulting system formed from the combination of the equalization filter and the room response should have a perceptually flat frequency response.

One of the important considerations is that the equalization filter has to be designed such that the spectral deviations in the magnitude response (e.g., Fig. 7.2) are minimized *simultaneously* for all listeners in the environment. Simultaneous equalization is an important consideration because listening has evolved into a group experience (e.g., as in home theaters, movie theaters, and concert halls). An example of performing only a single position equalization (by designing an inverse filter for position 1) is shown in Fig. 7.3. The top plot shows the equalization result at position

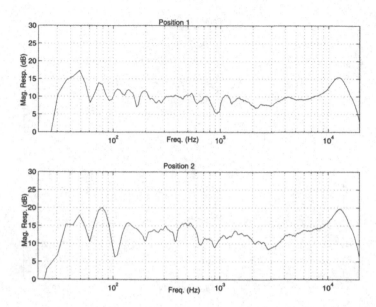

Fig. 7.2. Magnitude responses of room responses of Fig. 7.1 showing different spectral deviations (from flat) at the two listener positions.

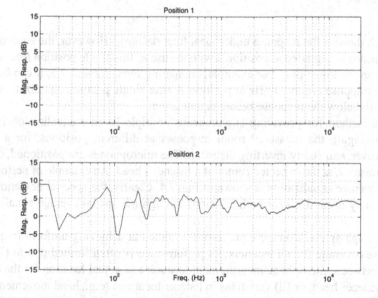

Fig. 7.3. Magnitude responses, upon single position equalization, of responses of Fig. 7.2. Specifically, the equalization filter is designed to correct for deviations at position 1, but the equalized response at position 2 is degraded.

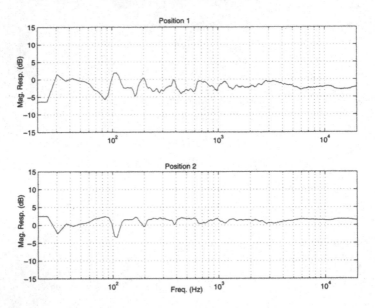

Fig. 7.4. Magnitude responses, upon spatial average equalization, of responses of Fig. 7.2. Specifically, the equalization filter is designed to correct for deviations, on an average, at positions 1 and 2.

1 (which shows a flat response under ideal filter design).[1] However, the equalization performance is degraded at position 2 with the use of this single position filter as can be seen in the lower plot. For example, comparing Figs. 7.2 and 7.3, it can be seen that the response around 50 Hz at position 2, after single position equalization, is at least 7 dB below the response before equalization.

One method for providing simultaneous multiple listener equalization is spatially averaging the measured room responses at different positions, for a given loudspeaker, and stably inverting the result. The microphones are positioned, during measurements, at the expected center of a listener's head. An example of performing spatial average equalization is shown in Fig. 7.4. Clearly, the spectral deviations are significantly minimized for both positions through the spatial average equalization filter.[2]

Although spatial average equalization is aimed at achieving uniform frequency response coverage for all listeners, its performance is often limited due to (i) mismatch between microphone measurement location and actual location for the center of the listener head, or (ii) variations in listener locations (e.g., head movements).

In this chapter, we present a method for evaluating the robustness of spatial averaging-based equalization, due to the introduction of variations in room responses

---

[1] In practice, a low-pass filter with a large cutoff frequency (e.g., 10 kHz), depending on the the direct to reverberant energy, is applied to the equalization filter to prevent audio from sounding bright.

[2] The filter was a finite impulse response filter of duration 8192 samples.

(generated either through (i) or (ii)), for rectangular listener arrangements relative to a fixed sound source. The proposed approach uses a statistical description for the reverberant field in the responses (viz., via the normalized correlation functions) in a rectangular listener configuration for a rectangular room.[3] A similar approach is followed in [70] for determining variations in performance. However, this was done with a single position equalization in mind and is focused for microphone array applications (e.g., sound source localization). Talantzis and Ward [99] used a similar analysis for understanding the effect of source displacements, but this analysis was also presented for a microphone array setup without spatial average equalization.

The advantage of the proposed approach is that (i) it is based on established theory of the statistical nature of reverberant sound fields [16]; (ii) it can be applied to a large frequency range above the Schroeder frequency, for typical size rooms, unlike modal equations which are valid for low frequencies having wavelengths greater $1/3 \min[L_x, L_y, L_z]$ [12]; and (iii) the computational complexity, due to the approximations, is low.

In the next section we introduce background necessary for the development of the robustness analysis. Specifically, an introduction is provided for the deterministic direct component, and the statistical reverberant field correlations. Subsequently we present the mismatch measure for analyzing the effects of mismatch between microphone (during measurement of room responses) and listener position (during playback) with a spatial average equalizer. Additionally, convergence analysis of the equalization mismatch error, for spatial average equalization, is presented at the end of the section. Results based on simulations for typical rectangular listener arrangements relative to a fixed source are presented for a rectangular configuration as this is fairly common in large environments (e.g., movie theaters, concert halls) as well as in typical home theater setups. The analysis can be extended to arbitrary listening configurations.

7.2 Room Acoustics for Simple Sources

The sound pressure $p_{f,i}$ at location i and frequency f can be expressed as a sum of the direct field component, $p_{f,d,i}$, and a reverberant field component, $p_{f,rev,i}$, as given by

$$p_{f,i} = p_{f,d,i} + p_{f,rev,i} \qquad (7.1)$$

The direct field component for sound pressure, $p_{f,d,i}$, of a plane wave, at far field listener location i for a sound source of frequency f located at i_0 can be expressed as [12]

$$p_{f,d,i} = -jk\rho c S_f g_f(i|i_0)e^{-j\omega t}$$
$$g_f(i|i_0) = \frac{1}{4\pi R}e^{jkR}$$

---

[3] A rectangular room is considered because the assumptions for a statistical reverberant sound field have been verified for this shape room and in practice rectangular-shaped rooms are commonly found.

$$R^2 = |i - i_0|^2 \tag{7.2}$$

where $p_{f,d}(i|i_0)$ is the direct component sound pressure amplitude, S_f is the source strength, $k = 2\pi/\lambda$ is the wavenumber, $c = \lambda f$ is the speed of sound (343 m/s) and ρ is the density of the medium (1.25 kg/m^3 at sea level).

The normalized correlation function [100] which expresses a statistical relation between sound pressures, of reverberant components, at separate locations i and j, is given by

$$\frac{E\{p_{f,rev,i}p^*_{f,rev,i}\}}{\sqrt{E\{p_{f,rev,i}p^*_{f,rev,i}\}}\sqrt{E\{p_{f,rev,j}p^*_{f,rev,j}\}}} = \frac{\sin kR_{ij}}{kR_{ij}} \tag{7.3}$$

where R_{ij} is the separation between the two locations i and j relative to an origin, and $E\{.\}$ is the expectation operator.

The reverberant field mean square pressure is defined as

$$E\{p_{f,rev,i}p^*_{f,rev,i}\} = \frac{4c\rho\Pi_a(1 - \bar{\alpha})}{S\bar{\alpha}} \tag{7.4}$$

where Π_a is the power of the acoustic source, $\bar{\alpha}$ is the average absorption coefficient of the surfaces in the room, and S is the surface area of the room.

The assumption of a statistical description (as given in (7.3), (7.4)) for reverberant fields in rooms is justified if the following conditions are fulfilled [16]. (1) Linear dimensions of the room must be large relative to the wavelength. (2) Average spacing of the resonance frequencies must be smaller than one-third of their bandwidth (this condition is fulfilled in rectangular rooms at frequencies above the Schroeder frequency, $f_s = 2000\sqrt{T_{60}/V}$ Hz (T_{60} is the reverberation time in seconds, and V is the volume in m^3). (3) Both source and microphone are in the interior of the room, at least a half-wavelength away from the walls.

Furthermore, under the conditions in [16], the direct and reverberant sound pressures are uncorrelated.

7.3 Mismatch Analysis for Spatial Average Equalization

7.3.1 Analytic Expression for Mismatch Performance Function

A performance function, W_f, that is used for analyzing the effects of mismatch, for spatial average equalization, of room responses is given as

$$\bar{W}_f = \frac{1}{N}\sum_{i=1}^{N}\epsilon_{f,i}(r)$$

$$\epsilon_{f,i}(r) = E\{|\tilde{p}_f(r)\bar{p}_f^{-1} - p_{f,i}\bar{p}_f^{-1}|^2\} \tag{7.5}$$

In (7.5), $\epsilon_{f,i}(r)$ represents the equalization error in the r-neighborhood of the equalized location i having response $p_{f,i}$ (r neighborhood is defined as all points at a distance of r from location i). The neighboring response, at a distance r from location i, is denoted by $\tilde{p}_f(r)$, whereas the spatial average equalization response is denoted by $\bar{p}_f$. Thus, response $\tilde{p}_f(r)$ is the response corresponding to the displaced center of head position of the listener (viz., with a displacement of r). To get an intermediate equalization error measure, $\epsilon_{f,i}(r)$, the expectation is performed over all neighboring locations at a distance r from the equalized location i. Furthermore, the final performance function $\bar{W}_f$ is the average of all the equalization errors, $\epsilon_{f,i}(r)$, in the vicinity of the N equalized locations. In essence, the displacement (distance) r can be interpreted as a "mismatch parameter," because a room response measured at displacement r will be different from the response measured at a nominal location i.

For simplicity, in our analysis, we assume variations in responses due to displacements (or mismatch) in a horizontal plane (i.e., the x and y plane). The analysis, presented in this chapter, can be extended to include displacements on a spherical surface. Thus, (7.5) can be simplified to yield

$$\epsilon_{f,i}(r) = E\left\{\left|\frac{\tilde{p}_f(r)N}{\sum_{j=1}^{N} p_{f,j}} - \frac{p_{f,i}N}{\sum_{j=1}^{N} p_{f,j}}\right|^2\right\} \tag{7.6}$$

An approximate simplification for (7.5) can be done by using the Taylor series expansion [101]. Accordingly, if g is a function of random variables, x_i, with average values $E\{x_i\} = \bar{x}_i$, then $g(x_1, x_2, \ldots, x_n) = g(x)$ can be expressed as $g(x) = g(\bar{x}) + \sum_{i=1}^{n} g_i'(\bar{x})(x_i - \bar{x}_i) + g(\hat{x})$, where $g(\hat{x})$ is a function of order 2 (i.e., all its partial derivatives up to the first-order vanish at $(\bar{x}_1, \bar{x}_2, \ldots, \bar{x}_n)$. Thus, to a zeroth-order of approximation $E\{g(x)\} \approx g(\bar{x})$.

Hence, an approximation for (7.6) is given as

$$\epsilon_{f,i}(r) \approx N^2 \frac{E\{\tilde{p}_f(r)\tilde{p}_f(r)^* - \tilde{p}_f(r)p_{f,i}^* - \tilde{p}_f(r)^* p_{f,i} + p_{f,i}p_{f,i}^*\}}{\sum_j \sum_k E\{p_{f,j}p_{f,k}^*\}} \tag{7.7}$$

We use the following identities for determining the denominator of (7.7).

$$E\{p_{f,j}p_{f,k}^*\} = E\{p_{f,d,j}p_{f,d,k}^* + p_{f,rev,j}p_{f,rev,k}^*\} \tag{7.8}$$

$$|kc\rho S_f|^2 = 4\pi\Pi_a c\rho \tag{7.9}$$

$$E\{p_{f,d,j}p_{f,d,k}^*\} = \frac{\Pi_a c\rho}{4\pi R_j R_k}e^{jk(R_j - R_k)} \tag{7.10}$$

$$E\{p_{f,rev,j}p_{f,rev,k}^*\} = \frac{4c\rho\Pi_a(1 - \bar{\alpha})}{S\bar{\alpha}}\frac{\sin kR_{jk}}{kR_{jk}} \tag{7.11}$$

$$R_{jk} = \sqrt{R_j^2 + R_k^2 - 2R_j R_k \cos\theta_{jk}} \tag{7.12}$$

In summary (7.8) is obtained by using (7.1) and knowing that the reverberant and direct field components of sound pressure are uncorrelated, (7.9) is derived in [12,

p. 311], (7.10) is determined by using (7.2) and (7.9), and (7.11) is determined from (7.3) and (7.4). In (7.12), which is the cosine law, θ_{jk} is the angle, subtended at the source at i_0, between locations j and k.

Thus, the denominator term in (7.7) is

$$
\sum_j \sum_k E\{p_{f,j} p_{f,k}^*\}
$$

$$
= \sum_j \sum_k \left(\frac{\Pi_a c \rho}{4\pi R_j R_k} e^{jk(R_j - R_k)} + \frac{4c\rho \Pi_a (1 - \bar{\alpha})}{S\bar{\alpha}} \frac{\sin k R_{jk}}{k R_{jk}} \right) \tag{7.13}
$$

Now, the first numerator term in (7.7) is

$$
E\{\tilde{p}_f(r) \tilde{p}_f(r)^*\} = E\{\tilde{p}_{f,d}(r) \tilde{p}_{f,d}(r)^* + \tilde{p}_{f,rev}(r) \tilde{p}_{f,rev}(r)^*\}
$$
$$
E\{\tilde{p}_{f,d} \tilde{p}_{f,d}^*\} = |k\rho c S_f|^2 E\{g_f(\tilde{i}|i_0) g_f^*(\tilde{i}|i_0)\}
$$
$$
= |k\rho c S_f|^2 E\left\{ \frac{1}{(4\pi)^2 |\tilde{R}|^2} \right\} \tag{7.14}
$$

where $\tilde{R}$ is the distance from a source at i_0 relative to spatial average equalized location i, and is determined by using cosine law (viz., $\tilde{R} = \sqrt{R_i^2 + r^2 - 2R_i r \cos \theta_i}$, where θ_i is the angle subtended at the source between location i and the location in the r-neighborhood of location i). The result from applying the expectation can be found by averaging over all locations in a circle in the r-neighborhood of location i (because for simplicity we have assumed mismatch in the horizontal or the x and y plane). Thus,

$$
E\left\{ \frac{1}{|4\pi \tilde{R}|^2} \right\} = \frac{1}{2} \frac{1}{(4\pi)^2} \int_{-1}^{1} \frac{d(\cos \theta_i)}{R_i^2 + r^2 - 2R_i r \cos \theta_i} \tag{7.15}
$$

Simplifying (7.15) and substituting the result in (7.14) gives

$$
E\{\tilde{p}_{f,d}(r) \tilde{p}_{f,d}(r)^*\} = \frac{|k\rho c S_f|^2}{2(4\pi)^2 R_i r} \log \left| \frac{R_i + r}{R_i - r} \right| = \frac{\Pi_a \rho c}{8 R_i r \pi} \log \left| \frac{R_i + r}{R_i - r} \right| \tag{7.16}
$$

$$
E\{\tilde{p}_{f,rev}(r) \tilde{p}_{f,rev}(r)^*\} = \frac{4c\rho \Pi_a (1 - \bar{\alpha})}{S\bar{\alpha}} \tag{7.17}
$$

The result in (7.16) is obtained by using (7.9), whereas (7.17) is a restatement of (7.4). Thus,

$$
E\{\tilde{p}_f(r) \tilde{p}_f(r)^*\} = \frac{\Pi_a \rho c}{8 R_i r \pi} \log \left| \frac{R_i + r}{R_i - r} \right| + \frac{4c\rho \Pi_a (1 - \bar{\alpha})}{S\bar{\alpha}} \tag{7.18}
$$

The correlation, $E\{\tilde{p}_{f,d}(r) p_{f,d,i}(r)^*\}$, in the direct-field component for the second term in the numerator of (7.17) is

$$
|k\rho c S_f|^2 \frac{1}{2(4\pi)^2} \int_{-1}^{1} \frac{e^{jk(\sqrt{R_i^2 + r^2 - 2R_i r \cos \theta_i} - R_i)} d \cos \theta_i}{R_i \sqrt{R_i^2 + r^2 - 2R_i r \cos \theta_i}} = \frac{\Pi_a \rho c}{4\pi R_i^2} \frac{1}{(4\pi R_i)^2} \frac{\sin kr}{kr}
$$
$$
\tag{7.19}
$$

The reverberant field correlation for the second term in the numerator of (7.7) can be found using (7.3), and is

$$E\{\tilde{p}_{f,rev}(r)p_{f,rev,i}^*\} = \frac{4c\rho\Pi_a(1-\bar{\alpha})}{S\bar{\alpha}}\frac{\sin kr}{kr} \tag{7.20}$$

The third numerator term in (7.7) can be found in a similar manner as compared to the derivation for (7.19) and (7.20).

The last term in the numerator of (7.7) is computed to yield

$$E\{p_{f,i}p_{f,i}^*\} = \frac{\Pi_a\rho c}{4\pi R_i^2} + \frac{4\rho c\Pi_a(1-\bar{\alpha})}{S\bar{\alpha}} \tag{7.21}$$

Equation (7.21) can be obtained by substituting $j = k = i$ in (7.10) and (7.11), respectively. Substituting the computed results into (7.7), and simplifying by canceling certain common terms in the numerator and the denominator, the resulting equalization error due to displacements (viz., mismatch in responses) is

$$\epsilon_{f,i}(r) \approx \frac{N^2}{\psi_1}\left[\frac{1}{8R_ir\pi}\log\left|\frac{R_i+r}{R_i-r}\right| + 2\psi_2 + \frac{1}{2\psi_3} - \left(\frac{1}{\psi_3} + 2\psi_2\right)\frac{\sin kr}{kr}\right] \tag{7.22}$$

$$\psi_1 = \sum_j\sum_l\left(\frac{1}{4\pi R_j R_l}e^{jk(R_j-R_l)} + \psi_2\frac{\sin kR_{jl}}{kR_{jl}}\right)$$

$$\psi_2 = \frac{4(1-\bar{\alpha})}{S\bar{\alpha}}$$

$$\psi_3 = 2\pi R_i^2$$

$$R_{jl} = \sqrt{R_j^2 + R_l^2 - 2R_j R_l\cos\theta_{jl}}$$

Finally, substituting (7.22) into (7.5) yields the necessary equation for $\bar{W}_f$.

7.3.2 Analysis of Equalization Error

In this section, we present an analysis of the behavior of the equalization error at each listener. This analysis helps in understanding, theoretically, the degradation (from a "steady-state" perspective) of equalization performance at different listener positions and at different frequencies.

Throughout the analysis we assume that $r < R_i$, for small mismatch between microphone position and center of listener head position relative to the distance between the microphone and the source. Thus, in (7.22) $\log|(R_i + r)/(R_i - r)| \to 0$.

Now, for $r/\lambda > 1$, the equalization error (7.22) converges to a steady-state value, $\epsilon_{f,i}^{ss}(r)$:

$$\epsilon_{f,i}^{ss}(r) \approx \frac{N^2}{\psi_1}\left[2\psi_2 + \frac{1}{2\psi_3}\right] = k_1\left(k_2 + \frac{1}{4\pi R_i^2}\right) \propto \frac{1}{R_i^2} \tag{7.23}$$

because $\sin kr/kr \to 0$. This implies that listeners at larger distances will have lower steady-state equalization errors than listeners closer to the source for a given wavelength of sound. Primarily, the inverse relationship between $\epsilon_{f,i}^{ss}(r)$ and R_i, at steady-state in (7.23), is due to the direct path sound field correlations (viz., $1/2\psi_3$ obtained from (7.21)) at position i.

7.4 Results

We simulated Eq. (7.22) for frequencies above the Schroeder frequency $f_s = 77$ Hz (i.e., $T_{60} = 0.7$ sec, $V = 8$ m $\times 8$ m $\times 8$ m).

In this setup, we simulated a rectangular arrangement of six microphones, with a source in the front of the arrangement. Specifically, microphones 1 and 3 were at a distance of 3 m from the source, microphone 2 was at 2.121 m, microphones 4 and 6 were at 4.743 m, and microphone 5 was at 4.242 m. The angles θ_{1k} in (7.12) were $(45, 90, 18.5, 45, 71.62)$ degrees for $(k = 2, \dots, 6)$, respectively. Thus, the distances of the listeners from the source are such that $R_6 = R_4 > R_5 > R_1 = R_3 > R_2$.

The equalization error, $\epsilon_{f,i}(r)$, results are depicted for different listeners in Figs. 7.5 to 7.8 for four frequencies ($f = 500$ Hz, $f = 1$ kHz, $f = 5$ kHz, and $f = 10$ kHz) as a function of r/λ, where the mismatch parameter $0 \leq r \leq 0.7$ m (r/λ corresponds to no mismatch condition). Specifically, only the results for listeners 1 and 2 are shown in the top panels because listener 3 is symmetric relative to source/listener 2 (hence the results of listener 1 and 3 are identical). Similarly, only the results for listeners 4 and 5 are shown in the bottom panels.

We observe the following.

1. It can be seen that the steady-state equalization error at listener 2 is higher than that at listener 1 (top panel). This follows from Eq. (7.23) (because $R_1 = R_3 > R_2$). Similar results can be predicted for the equalization errors for listeners 4 and 5 (this is not immediately obvious in the bottom panels, because R_4 is close to R_5).

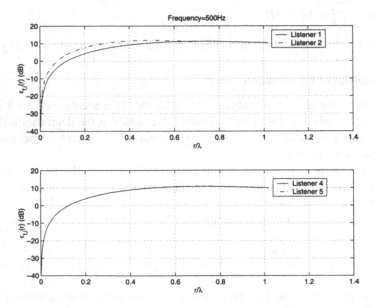

Fig. 7.5. $\epsilon_{f,i}(r)$ for the listeners at different distances from the source, $r/\lambda = 0$ corresponds to the optimal position, $f = 500$ Hz.

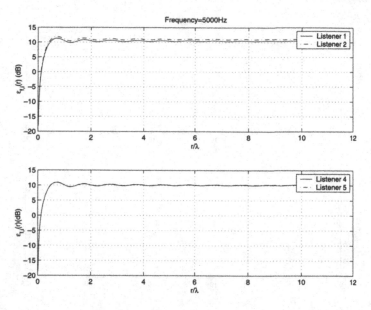

Fig. 7.6. $\epsilon_{f,i}(r)$ for the listeners at different distances from the source, $f = 1$ kHz.

2. Furthermore, the nonsteady-state equalization region, for a given equalization error, is larger (better) for listeners farther from the source. For example, the equalization region is a circle of a radius 0.025λ for listener 2, whereas it is 0.04λ for listener 1 at $\epsilon_{f,i}(r) = -10$ dB and $f = 500$ Hz. This effect is dominant at lower

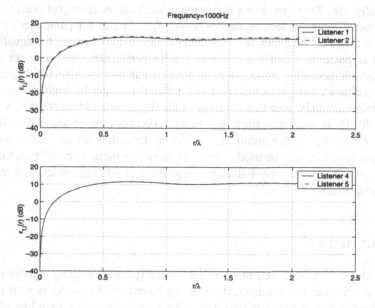

Fig. 7.7. $\epsilon_{f,i}(r)$ for the listeners at different distances from the source, $f = 5$ kHz.

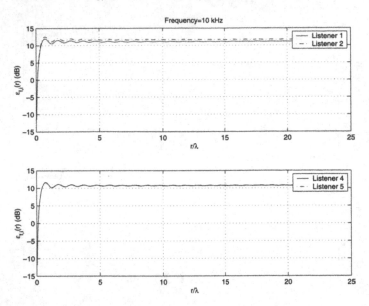

Fig. 7.8. $\epsilon_{f,i}(r)$ for the listeners at different distances from the source, $f = 10$ kHz.

frequencies but not easily noticeable at higher frequencies (as can be seen from the initial rise of the error towards a peak value before reaching a steady-state value).

3. The equalization error shows a sinc$(2r/\lambda)$ dependance after the initial peak (as emphasized in Fig. 7.6). This dependence arises from the finite correlation of the reverberant field before it reaches a negligible value at steady-state.

Finally, Fig. 7.9 summarizes the average equalization error plot (i.e., W_{avg} of (7.5)), over all listeners, for frequencies beyond f_s and mismatch parameter r ranging from 0 m to 0.7 m. This measure is a composite measure weighing the equalization error at all positions equally, and shows that the performance degrades for all frequencies with increasing mismatch or displacement in meters. Also, the degradation, for small displacement r (of the order of 0.1 meter) is larger for higher frequencies. For example, it can be seen that the slope of the W_{avg} curves in the frequency region around 200 Hz is lower than the slopes of the curves for frequencies around 10 kHz. Alternate measures with nonuniform weighting, depending on the "importance" of a listening position, may be used instead. Thus, such a measure could potentially be used to give an overall picture during comparisons to other approaches of multiple listener equalization.

7.5 Summary

In this chapter, we analyzed the performance of spatial averaging equalization, in a multiple listener environment, used during sound playback. As is well known, room equalization at multiple positions, allows for high-quality sound playback in

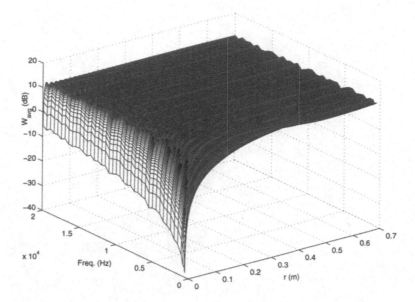

Fig. 7.9. $\bar{W}$ for various mismatch parameters and frequencies between 20 Hz and 20 kHz.

the room. However, as is typically the case in room equalization, the microphone positions during measurement of the room response will not necessarily correspond to the center of head of the listener leading to a frequency-dependent degradation due to mismatch between the measured response and the actual response corresponding to the center of listener head during playback. Several interesting observations can be made from the results, including: (i) the influence of frequency and distance on the size of equalization region, (ii) the steady-state equalization error being dependent on the distance of the listener from the source, and (iii) the dependence of the reverberant field correlation on the equalization error. Future goals can be directed to using the proposed method for comparing different multiple listener equalization techniques in terms of their robustness to response mismatch.

8

Robustness of Equalization to Displacement Effects: Part II

In a multiple listener environment, equalization may be performed through magnitude response spatial averaging at expected listener positions. However, the performance of averaging-based equalization, at the listeners, will be affected when there is a mismatch between microphone and listener positions. In this chapter, we present a modal analysis approach, targeted at low frequencies, to map mismatch to an equalization performance metric. Specifically, a closed-form expression is provided that predicts the equalization performance in the presence of mismatch. The results, which are particularly valid at lower frequencies where standing wave modes of the room are dominant, indicate that magnitude average equalization performance depends on (i) the amount of displacement/mismatch, and (ii) the frequency component in the modal response. We have provided validation of the theoretical results, thereby indicating the usefulness of the proposed analytic approach for measuring equalization performance due to mismatch effects. We also demonstrate the importance of average equalization over single listener equalization when considering mismatch/displacement effects.

8.1 Introduction

In this chapter, we propose a statistical approach, using modal equations, for evaluating the robustness of magnitude response average equalization, due to variations in room responses (due to (i) or (ii)). Specific results are obtained for a particular listener arrangement in a rectangular room. Modal equations have been used in the analysis because they accurately model the magnitude response at low frequencies. As is well known, dominant modes, in this low-frequency region, are relatively harder to equalize than at higher frequencies. In the next section, we introduce the necessary background used in the development of the proposed robustness analysis. The subsequent section is devoted to the development of the robustness analysis for spatial average-based equalization. Results based on simulations for a typical rectangular listener arrangement relative to a fixed source and validation of the theoretical analysis are presented.

8.2 Modal Equations for Room Acoustics

The Green's function derived from the wave theory for sound fields in an enclosure is given by [11, 12] as

$$p_\omega(\underline{q}_l) = jQ\omega\rho_0 \sum_{\underline{n}} \frac{p_{\underline{n}}(\underline{q}_l)p_{\underline{n}}(\underline{q}_o)}{K_{\underline{n}}(k^2 - k_{\underline{n}}^2)} = jQ\omega\rho_0 \sum_{n_x=0}^{N_x-1}\sum_{n_y=0}^{N_y-1}\sum_{n_z=0}^{N_z-1} \frac{p_{\underline{n}}(\underline{q}_l)p_{\underline{n}}(\underline{q}_o)}{K_{\underline{n}}(k^2 - k_{\underline{n}}^2)}$$

$$\underline{n} = (n_x, n_y, n_z); \ k = \omega/345; \ \underline{q}_l = (x_l, y_l, z_l)$$

$$k_{\underline{n}} = \pi\left[\left(\frac{n_x}{L_x}\right)^2 + \left(\frac{n_y}{L_y}\right)^2 + \left(\frac{n_z}{L_z}\right)^2\right]^{1/2} \int\int_V\int p_{\underline{n}}(\underline{q}_l)p_{\underline{m}}(\underline{q}_l)dV$$

$$= \begin{cases} K_{\underline{n}} & \underline{n} = \underline{m} \\ 0 & \underline{n} \neq \underline{m} \end{cases} \tag{8.1}$$

where the eigenfunctions $p_{\underline{n}}(\underline{q}_l)$ can be assumed to be orthogonal to each other under certain conditions, with the point source being at $\underline{q}_o$. The modal equations in (8.1) are valid for wavelengths, λ, where $\lambda > (1/3)\min[L_x, L_y, L_z]$ [12]. At these low frequencies, a few standing waves are excited, so that the series terms in (8.1) converge quickly.

For a rectangular enclosure with dimensions (L_x, L_y, L_z), $\underline{q}_o = (0, 0, 0)$, the eigenfunctions $p_{\underline{n}}(\underline{q}_l)$ and eigenvalues K_n in (8.1) are

$$p_{\underline{n}}(\underline{q}_l) = \cos\left(\frac{n_x\pi x_l}{L_x}\right)\cos\left(\frac{n_y\pi y_l}{L_y}\right)\cos\left(\frac{n_z\pi z_l}{L_z}\right)$$

$$p_{\underline{n}}(\underline{q}_o) = 1$$

$$K_{\underline{n}} = \int_0^{L_x}\cos^2\left(\frac{n_x\pi x_l}{L_x}\right)dx\int_0^{L_y}\cos^2\left(\frac{n_y\pi y_l}{L_y}\right)dy\int_0^{L_z}\cos^2\left(\frac{n_z\pi z_l}{L_z}\right)dz$$

$$= \frac{L_xL_yL_z}{8} = \frac{V}{8} \tag{8.2}$$

The eigenfunction distribution in the $z = 0$ plane, for a room of dimension 6 m ×6 m ×6 m, and tangential mode $(n_x, n_y, n_z) = (3, 2, 0)$ is shown in Fig. 8.1. Thus, the large deviation in the eigenfunction distribution, for different modes in the room, necessitates a multiple listener room equalization method.

8.3 Mismatch Analysis with Spatial Average Equalization

8.3.1 Spatial Averaging for Multiple Listener Equalization

The magnitude response averaging method is popular for performing spatial equalization over a wide area in a room. The magnitude response spatial averaging process can be expressed in terms of the modal equations (8.1) as

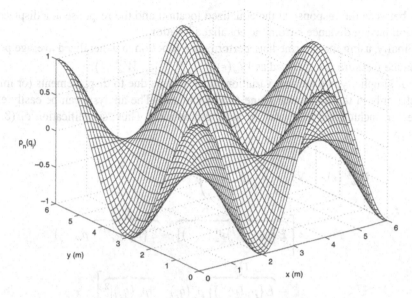

Fig. 8.1. The eigenfunction distribution, for a tangential mode (3,2,0) in the $z = 0$ plane, over a room of dimensions 6 m ×6 m ×6 m.

$$p_{\omega,avg} = \left(\frac{1}{N} \sum_{l=1}^{N} |p_\omega(\underline{q}_l)| \right) \tag{8.3}$$

where N is the number of positions that are to be equalized in the room.

The spatial equalizer, $p_{\omega,avg}^{-1}$, filters the audio signal before it is transmitted by a loudspeaker in the room. A block diagram of the multiple listener equalization process, using averaging, is shown in Fig. 8.2.

8.3.2 Equalization Performance Due to Mismatch

An intermediate performance function, $W_\omega^{(i)}(\epsilon)$, that is used for analyzing the robustness of spatial average equalization to room response variations is given as

$$W_\omega^{(i)}(\epsilon) = E\{|p_\omega(\underline{\nu}_\epsilon^{(i)})p_{\omega,avg}^{-1} - p_\omega(\underline{q}_i)p_{\omega,avg}^{-1}|^2\} \tag{8.4}$$

where $p_\omega(\underline{\nu}_\epsilon^{(i)})$ is the pressure at location $\underline{\nu}$ in the ϵ neighborhood of equalized position i having pressure $p_\omega(\underline{q}_i)$ (ϵ-neighborhood is defined as all points at a distance of ϵ from location i), $E\{.\}$ denotes the statistical expectation operator, and $\omega = 2\pi c/\lambda$ (where $c = 345m/s$).

The intermediate performance measure in (8.4) is defined in such a manner that when the displacement ϵ, about position i (whose response $p_\omega(\underline{q}_i)$ which is originally used for determining the spatially averaged equalization filter $p_{\omega,avg}^{-1}$) is zero, then $W_\omega^{(i)}(\epsilon) = 0$. Thus, the performance measure is computed as an average square error between the response at the equalized location and the response at a displaced location having distance ϵ from the equalized location.

Finally, using the intermediate performance function, a generalized average performance measure is expressed as $W_\omega(\epsilon) = 1/N \sum_{i=1}^{N} W_\omega^{(i)}(\epsilon)$.

For simplicity, we assume variations in responses due to displacements (or mismatch) only in the horizontal plane (x and y plane). The analysis can be easily extended to include the mismatch in three dimensions. Thus, simplification of (8.4) leads to

$$W_\omega^{(i)}(\epsilon) = N^2 / \left(\sum_{l=1}^{N} |p_\omega(\underline{q}_l)| \right)^2$$

$$\times \left[\overbrace{E\{p_\omega(\underline{\nu}_\epsilon^{(i)})p_\omega^*(\underline{\nu}_\epsilon^{(i)})\}}^{I} - \overbrace{E\{p_\omega^*(\underline{\nu}_\epsilon^{(i)})\}}^{II} p_\omega(\underline{q}_i) \right.$$

$$\left. - \overbrace{E\{p_\omega(\underline{\nu}_\epsilon^{(i)})\}}^{III} p_\omega^*(\underline{q}_i) + \overbrace{|p_\omega(\underline{q}_i)|^2}^{IV} \right] \tag{8.5}$$

We only need to compute the statistics associated with Terms (I), (II), and (III) (the terms within the expectations) in (8.5), because Term (IV) is a deterministic quantity.

Now, $E\{p_\omega(\underline{\nu}_\epsilon^{(i)})p_\omega^*(\underline{\nu}_\epsilon^{(i)})\}$ is the average over all locations along a circle of radius ϵ from the ith listener location. Assuming the source, all listeners, and each of the listener displacements are along the same z-plane (viz., $z = 0$), then (I) in (8.5) can be simplified as

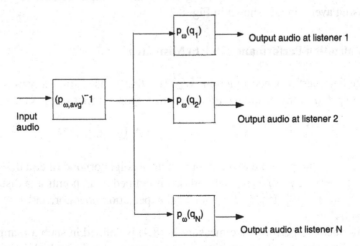

Fig. 8.2. Spatial average equalization for N listeners in a room.

$$p_\omega(\underline{\nu}_\epsilon^{(i)}) = \frac{j8Q\omega\rho_0}{V} \sum_{\underline{n}} \frac{\cos\left(\frac{n_x\pi\phi_x^{(i)}}{L_x}\right)\cos\left(\frac{n_y\pi\phi_y^{(i)}}{L_y}\right)}{(k^2 - k_{\underline{n}}^2)} \tag{8.6}$$

$$E\{p_\omega(\underline{\nu}_\epsilon^{(i)})p_\omega^*(\underline{\nu}_\epsilon^{(i)})\} = |\psi_1|^2 \sum_{n,m} (1/\psi_2)\psi_3 \tag{8.7}$$

$$\psi_1 = \frac{8Q\omega\rho_0}{V}$$

$$\psi_2 = (k^2 - k_{\underline{n}}^2)(k^2 - k_{\underline{m}}^2)$$

$$\psi_3 = E\left\{ \cos\left(\frac{n_x\pi\phi_x^{(i)}}{L_x}\right)\cos\left(\frac{n_y\pi\phi_y^{(i)}}{L_y}\right) \right.$$
$$\left. \times \cos\left(\frac{m_x\pi\phi_x^{(i)}}{L_x}\right)\cos\left(\frac{m_y\pi\phi_y^{(i)}}{L_y}\right) \right\} \tag{8.8}$$

Now, with $\phi_x^{(i)} = x_i + \epsilon\cos\theta$ and $\phi_y^{(i)} = y_i + \epsilon\sin\theta$

$$E\left\{ \cos\left(\frac{n_x\pi\phi_x^{(i)}}{L_x}\right)\cos\left(\frac{n_y\pi\phi_y^{(i)}}{L_y}\right)\cos\left(\frac{m_x\pi\phi_x^{(i)}}{L_x}\right)\cos\left(\frac{m_y\pi\phi_y^{(i)}}{L_y}\right) \right\}$$
$$= \frac{1}{2\pi}\int_0^{2\pi} \cos\left(\frac{n_x\pi(x_i + \epsilon\cos\theta)}{L_x}\right)\cos\left(\frac{n_y\pi(y_i + \epsilon\sin\theta)}{L_y}\right)$$
$$\times \cos\left(\frac{m_x\pi(x_i + \epsilon\cos\theta)}{L_x}\right)\cos\left(\frac{m_y\pi(y_i + \epsilon\sin\theta)}{L_y}\right) d\theta. \tag{8.9}$$

Equation (8.9) can be solved using the MATLAB *trapz* function. However, we found an approximate closed-form expression to be computationally much faster. The following expressions were derived from standard trigonometric formulae, using the first two terms in the polynomial expansion of the cosine function, and the first term in the polynomial expansion of the sine function because $(\epsilon/L_x, \epsilon/L_y, \epsilon/L_z) \ll 1$. Thus,

$$E\left\{ \cos\left(\frac{n_x\pi\phi_x^{(i)}}{L_x}\right)\cos\left(\frac{n_y\pi\phi_y^{(i)}}{L_y}\right)\cos\left(\frac{m_x\pi\phi_x^{(i)}}{L_x}\right)\cos\left(\frac{m_y\pi\phi_y^{(i)}}{L_y}\right) \right\}$$
$$= \frac{1}{2\pi}(A + B + C) \tag{8.10}$$

where

$$A = \pi\cos\left(\frac{n_x\pi x_i}{L_x}\right)\cos\left(\frac{n_y\pi y_i}{L_y}\right)\cos\left(\frac{m_x\pi x_i}{L_x}\right)\cos\left(\frac{m_y\pi y_i}{L_y}\right)$$
$$\times \left[2 - \epsilon_y^2 v_y - \epsilon_x^2 v_x + \frac{3}{4}(\epsilon_x^4 u_x^4 + \epsilon_y^4 u_y^4) - \frac{1}{8}(\epsilon_x^2\epsilon_y^4 u_y^2 v_x + \epsilon_x^4\epsilon_y^2 u_x^2 v_y) \right.$$
$$\left. + \frac{1}{4}\epsilon_x^2\epsilon_y^2 v_x v_y + \frac{3}{64}\epsilon_x^4\epsilon_y^4 u_x^2 u_y^2 \right]$$

$$B = \pi \epsilon_y^2 u_y \cos\left(\frac{n_x \pi x_i}{L_x}\right) \cos\left(\frac{m_x \pi x_i}{L_x}\right) \sin\left(\frac{n_y \pi y_i}{L_y}\right) \sin\left(\frac{m_y \pi y_i}{L_y}\right)$$

$$\times \left[2 - 0.5\epsilon_x^2 \left(m_x^2 + n_x^2 - 0.5\epsilon_x^2 u_x^2\right)\right]$$

$$C = \pi \epsilon_y^2 \epsilon_x^2 u_x u_y \sin\left(\frac{n_x \pi x_i}{L_x}\right) \sin\left(\frac{n_y \pi y_i}{L_y}\right) \sin\left(\frac{m_x \pi x_i}{L_x}\right) \sin\left(\frac{m_y \pi y_i}{L_y}\right)$$

$$+ \pi \epsilon_x^2 u_x \sin\left(\frac{n_x \pi x_i}{L_x}\right) \sin\left(\frac{m_x \pi x_i}{L_x}\right) \cos\left(\frac{m_y \pi y_i}{L_y}\right) \cos\left(\frac{n_y \pi y_i}{L_y}\right)$$

$$\times \left[2 - 0.5\epsilon_y^2 \left(n_y^2 + m_y^2 - 0.5\epsilon_y^2 u_y^2\right)\right] \tag{8.11}$$

where

$$\epsilon_x = \frac{\pi \epsilon}{\sqrt{2} L_x} \qquad\qquad \epsilon_y = \frac{\pi \epsilon}{\sqrt{2} L_y}$$

$$u_x = n_x m_x \qquad\qquad u_y = n_y m_y$$

$$v_x = (m_x^2 + n_x^2) \qquad v_y = (m_y^2 + n_y^2) \tag{8.12}$$

Thus (8.10) can be substituted in (8.7), and subsequently in (8.5), to determine Term I.

Now Terms (II) and (III) in (8.5) can be combined to give

$$-E\left\{p_\omega^*(\nu_\epsilon^{(i)})\right\} p_\omega(\underline{q}_i) - E\left\{p_\omega(\nu_\epsilon^{(i)})\right\} p_\omega^*(\underline{q}_i)$$

$$= -2\left|\frac{8Q\omega\rho_0}{V}\right|^2 \sum_{\underline{m},\underline{n}} \frac{p_{\underline{m}}(\underline{q}_i) E\left\{\cos\left(\frac{n_x \pi \phi_x^{(i)}}{L_x}\right) \cos\left(\frac{n_y \pi \phi_y^{(i)}}{L_y}\right)\right\}}{(k^2 - k_{\underline{m}}^2)(k^2 - k_{\underline{n}}^2)} \tag{8.13}$$

Again using $\phi_x^{(i)} = x_i + \epsilon \cos\theta$; $\phi_y^{(i)} = y_i + \epsilon \sin\theta$, we have

$$E\left\{\cos\left(\frac{n_x \pi \phi_x^{(i)}}{L_x}\right) \cos\left(\frac{n_y \pi \phi_y^{(i)}}{L_y}\right)\right\}$$

$$= \frac{1}{2\pi} \int_0^{2\pi} \cos\left(\frac{n_x \pi (x_i + \epsilon \cos\theta)}{L_x}\right) \cos\left(\frac{n_y \pi (y_i + \epsilon \sin\theta)}{L_y}\right) d\theta \tag{8.14}$$

Thus, upon again using the fact that $(\epsilon/L_x, \epsilon/L_y, \epsilon/L_z) \ll 1$, we can solve (8.14) as

$$E\left\{\cos\left(\frac{n_x \pi \phi_x^{(i)}}{L_x}\right) \cos\left(\frac{n_y \pi \phi_y^{(i)}}{L_y}\right)\right\}$$

$$= \frac{1}{2\pi} \cos\left(\frac{n_x \pi x_i}{L_x}\right) \cos\left(\frac{n_y \pi y_i}{L_y}\right) \left[2\pi - \pi \left(\epsilon_x^2 n_x^2 + \epsilon_y^2 n_y^2\right) + \frac{\pi}{4}\epsilon_y^2 \epsilon_x^2 n_x^2 n_y^2\right] \tag{8.15}$$

Substituting (8.15) in (8.13) and subsequently into (8.5) gives Terms II and III.

8.4 Results

8.4.1 Magnitude Response Spatial Averaging

Simulation of Eq. (8.5) was performed for a room of dimensions 6 m × 6 m × 6 m with six positions that were equalized by magnitude response averaging. The six listening position coordinates, relative to the source at (0 m, 0 m, 0 m), correspond to (−2.121 m, −2.121 m, 0 m); (0 m, −2.121 m, 0 m); (2.121 m, −2.121 m, 0 m); (−2.121 m, −4.242 m, 0 m); (0 m, −4.242 m, 0 m); (2.121 m, −4.242 m, 0 m). Figure 8.3 shows the configuration that was tested for six listener positions (depicted with asterisks) sitting in front of a source (depicted by a circle). Results are obtained for the lower 1/3 octave frequencies because equalization, in this region, is of greater importance (as large amplitude standing waves in this region are relatively difficult to correct). The frequencies used for the simulations were (25, 31.5, 40, 50, 63, 80, 125, 160, 200) Hz, where the lower 1/3 octave frequencies were 25 Hz, 31.5 Hz, 40 Hz; the middle 1/3 octave frequencies were 50 Hz, 63 Hz, 80 Hz; and the remainder comprising the upper 1/3 octave frequencies for the set of frequencies being considered.

Figure 8.4 shows the magnitude of the sound pressures, from 25 Hz up to 200 Hz at 1/3 octave frequencies, that can be expected for each of the listener positions. It should be noted that even though $f = 200$ Hz $\Rightarrow \lambda = 1.715$ m $< (1/3)(6) = 2$ m, we don't expect erroneous results at this frequency, because the wavelength at this frequency is not significantly offset from the wavelength limit imposed by the condition $\lambda > (1/3) \min[L_x, L_y, L_z]$. As can be seen, some listener positions exhibit

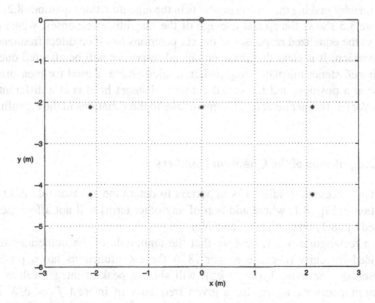

Fig. 8.3. Simulated setup for a six-position displacement effects analysis in a rectangular configuration in front of a source.

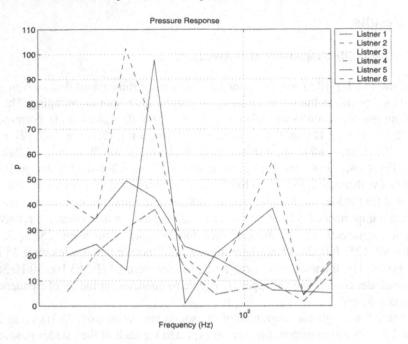

Fig. 8.4. Magnitude of the sound pressures, from 25 Hz up to 200 Hz at 1/3 octave frequencies, that can be expected for each of the listener positions.

the same response, because these positions are symmetrically located relative to the source, thereby making the cosine products in the eigenfunction equation (8.2) equal.

Figure 8.5 shows the spatial average of the magnitude responses, whereas Fig. 8.6 shows the equalized responses at the six positions based on direct frequency domain inversion. It is clear that these equalized responses will be affected due to (i) microphone/listener mismatch (e.g., when a microphone is used for measuring the response at a position, and the actual center of listener head is at a different position) and/or (ii) listener head displacement, due to the variations in the eigenfunction $p_{\underline{n}}(\underline{q}_l)$.

8.4.2 Computation of the Quantum Numbers

To get fairly accurate results, it is important to determine a reasonable limit of the summation in Eq. (8.1) where addition of any other terms will not affect the result generated by (8.1) (viz., N_x, N_y, and N_z).

For a rectangular room, observe that the amplitude of the numerator in (8.1) is bounded by unity (because as per (8.2) the eigenfunctions are a product of co-sinusoids). The term, $1/(k^2 - k_{\underline{n}}^2)$, will show a peak at integer values of the quantum numbers n_x, n_y, n_z for a given frequency of interest $f = c/\lambda$. Figure 8.7 shows a 3-D plot for the value of $1/(k^2 - k_{\underline{n}}^2)$ as a function of n_x, n_y for a 1/3 octave frequency of 63 Hz. From the figure it is clear that the summation,

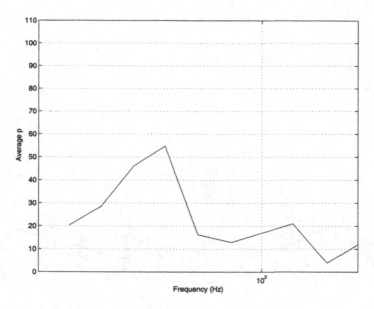

Fig. 8.5. Spatial average of the magnitude responses.

$\sum_{n_x=0}^{N_x-1} \sum_{n_y=0}^{N_y-1} p_{\underline{n}}(\underline{q_l})/(k^2 - k_{\underline{n}}^2)$, will be accurate if the first ten integers of n_x, n_y

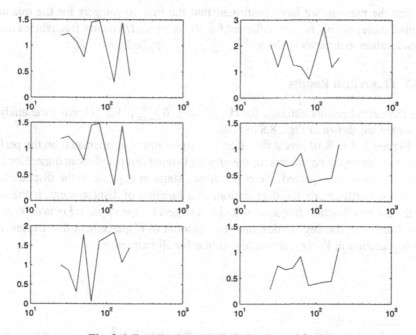

Fig. 8.6. Equalized responses at the six positions.

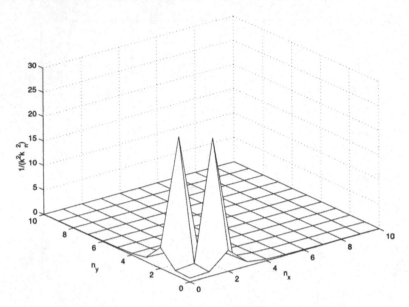

Fig. 8.7. Effect of quantum numbers on the summation of (8.1).

are used because, $1/(k^2 - k_{\underline{n}}^2) \approx 0$ for $n_x > 10$ and $n_y > 10$ for wavelength $\lambda = 5.4$ m.[1]

For the present, we have confirmed that the first 10 integers for the quantum number tuple, n_x, n_y, n_z, are sufficient for all the lower 1/3 octave frequencies under consideration to accurately model $\sum_{n_x} \sum_{n_y} \sum_{n_z} p_{\underline{n}}(\underline{q_l})/(k^2 - k_{\underline{n}}^2)$.

8.4.3 Theoretical Results

The theoretical results obtained for $W_\omega(\epsilon) = 1/6 \sum_{i=1}^{6} W_\omega^{(i)}(\epsilon)$ using the analysis presented are shown in Figs. 8.8 to 8.10.

Figures 8.8 to 8.10 reveal the effect of displacement or mismatch on the performance of averaging equalizers for the present listener setup and room dimensions. In every situation, as expected, the performance starts to degrade as the displacement increases. Furthermore, the degradation, as a function of displacement, is the least for the lower 1/3 octave frequencies and it is largest for the upper 1/3 octave frequencies. Moreover, the degradation rate (i.e., amount of displacement that causes a 10 dB degradation in $W_\omega(\epsilon)$) is roughly similar for all frequencies.

---

[1] Even though the plot shows significant attenuation of the $1/(k^2 - k_{\underline{n}}^2)$ term at $n_x > 6$ and $n_y > 6$, we have chosen the limit N_x and N_y, liberally, to account for the negligible, but nonzero, additional terms.

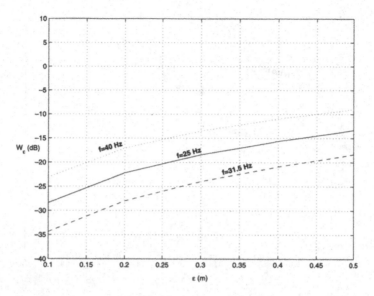

Fig. 8.8. $W_\omega(\epsilon)$ versus displacement ϵ for lower 1/3 octave frequencies.

8.4.4 Validation

This section validates the results obtained from using the closed-form expressions from the previous section. For this we did the following, for a given 1/3 octave frequency.

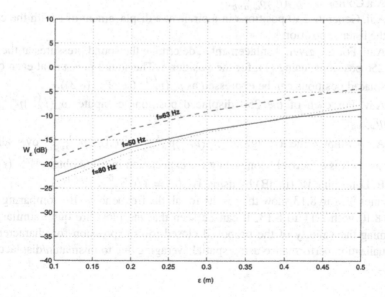

Fig. 8.9. $W_\omega(\epsilon)$ versus displacement ϵ for middle 1/3 octave frequencies.

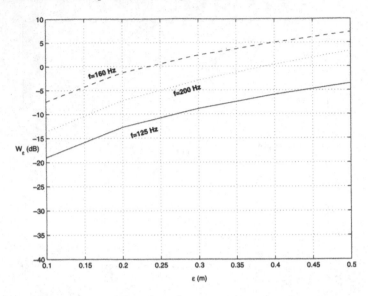

Fig. 8.10. $W_\omega(\epsilon)$ versus displacement ϵ for upper 1/3 octave frequencies.

1. Determine the sound pressures, $p_\omega(\underline{q}_i)$, using (8.1) for the six positions ($i = 1, 2, \ldots, 6$).
2. Determine the average of the sound pressure magnitudes at the six positions.
3. Determine the inverse of the average $p_{\omega,avg}^{-1}$.
   A. For each position i ($i = 1, 2, \ldots, 6$):
   A.i. Compute $-p_\omega(\underline{q}_i)p_{\omega,avg}^{-1}$.
   A.ii. Generate 250 positions in a circle at a displacement of ϵ with the center being the listener position i.
   A.iii. For the given displacement ϵ, determine the sound pressures at the each of the 250 positions, using (1), for the position i. The sound pressure at each of the 250 displaced positions can be expressed as $p_\omega(\nu_\epsilon^{(i)})$ (see Eq. (8.3)).
   A.iv. For each of the 250 displaced positions compute $|p_\omega(\nu_\epsilon^{(i)})p_{\omega,avg}^{-1} - p_\omega(\underline{q}_i)p_{\omega,avg}^{-1}|^2$
   A.v. Compute the average of $|p_\omega(\nu_\epsilon^{(i)})p_{\omega,avg}^{-1} - p_\omega(\underline{q}_i)p_{\omega,avg}^{-1}|^2$ over all 250 positions. This is effectively computing the expectation in (3) to obtain $W_\omega^{(i)}(\epsilon)$.
   B. Determine W (in dB) by using $W_\omega(\epsilon) = 1/6 \sum_{i=1}^{6} W_\omega^{(i)}(\epsilon)$.
Figures 8.11 to 8.13 show the results for all the frequencies. By comparing Figs. 8.8 to 8.10 with 8.11 to 8.13, it can be seen that the plots are quite similar, thus confirming the validity of the proposed closed-form expression for characterizing the equalization performance using spatial averaging due to mismatch/displacement effects.

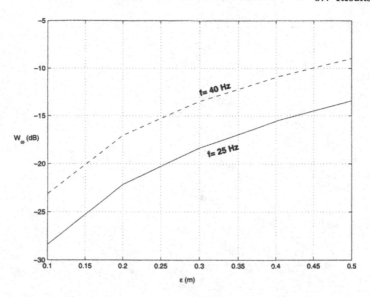

Fig. 8.11. Validation of the analytic solution for $W_\omega(\epsilon)$ versus displacement ϵ for lower 1/3 octave frequencies.

8.4.5 Magnitude Response Single-Listener Equalization

In this section we present some results obtained from single listener equalization. It is well known that single listener equalization may cause significant degradations,

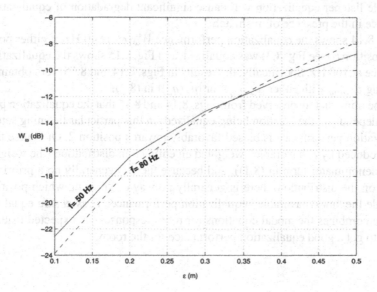

Fig. 8.12. Validation of the analytic solution for $W_\omega(\epsilon)$ versus displacement ϵ for middle 1/3 octave frequencies.

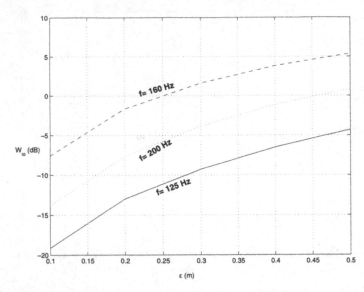

Fig. 8.13. Validation of the analytic solution for $W_\omega(\epsilon)$ versus displacement ϵ for upper 1/3 octave frequencies.

in the frequency response, at other listening positions. In fact, the degradation at the other positions could end up being more than what it would have been if the single position were not equalized [102]. Thus, the goal of this section is to further demonstrate that, besides degradation in frequency response at other listening positions, single listener equalization will cause significant degradation of equalization performance in the presence of mismatch.

Figure 8.14 shows the equalization performance $W_\omega(\epsilon)$, at 40 Hz, if either position 2 or position 6 (see Fig. 8.3) was equalized, and Fig. 8.15 shows the equalization performance at 160 Hz. Specifically, the results in Figs. 8.14 and 8.15 were obtained by replacing $p_{\omega,avg}$ with either $|p_\omega(\underline{q}_2)|$ or $|p_\omega(\underline{q}_6)|$ in (8.4).

It can be immediately observed from Figs. 8.14 and 8.15 that the equalization performance depends on the position being equalized. In this particular listening setup, the equalization performance is biased favorably towards position 2. Of course this bias is introduced by a "favorable" weighted eigenfunction distribution (the weights being the denominator term in (8.1)), and because there is generally no a priori information on the distribution, there is generally no way of knowing which position will provide the "most favorable" equalization performance. Thus, a safe equalization choice combines the modal equations (or room responses), at expected listener positions, to get a good equalization performance for the room.

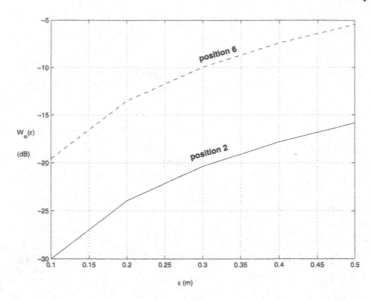

Fig. 8.14. Single listener equalization results for $f = 40$ Hz, where equalization is done for position 6 (dashed line) or position 2 (solid line).

8.5 Summary

In this chapter we presented a statistical approach using modal equations for evaluating the robustness of equalization based on magnitude response averaging, due to the variations in room responses, for a realistic listener arrangement relative to a source. The simulations were performed for a six-listener setup with a simple source in a cubic room.

Clearly, there is a degradation in the equalization performance due to displacement effects. Furthermore, this degradation is different for different frequencies (viz., generally smaller for relatively lower frequencies as compared to higher frequencies). We have also experimentally confirmed the validity of the proposed closed-form solution for measuring degradation performance.

Furthermore, we also demonstrated the importance of average equalization over single listener equalization when considering mismatch/displacement effects.

Finally, an interesting future research direction is the formulation of a perceptually motivated performance function and evaluation of the robustness using this measure.

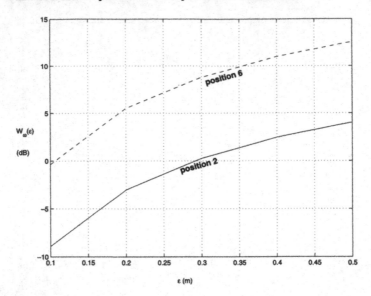

Fig. 8.15. Single listener equalization results for $f = 160$ Hz, where equalization is done for position 6 (dashed line) or position 2 (solid line).

9

Selective Audio Signal Cancellation

Selectively canceling signals at specific locations within an acoustical environment with multiple listeners is of significant importance for home theater, automobile, teleconferencing, office, industrial, and other applications. The traditional noise cancellation approach is impractical for such applications because it requires secondary sources to "anti-phase" the primary source, or sensors to be placed on the listeners. In this chapter we present an alternative method for signal cancellation by preprocessing the acoustical signal with a filter known as the eigenfilter [103, 104]. We examine the theoretical properties of such filters, and investigate the performance (gain) and tradeoff issues such as spectral distortion. Sensitivity of the performance as a function of the room impulse response duration (reverberation) modeled in the eigenfilter is also investigated.

9.1 Introduction

Integrated media systems are envisioned to have a significant impact on the way media, such as audio, are transmitted to people in remote locations. In media applications, although a great deal of ongoing research has focused on the problem of delivering high-quality audio to a listener, the problem of delivering appropriate audio signals to multiple listeners in the same environment has not yet been adequately addressed.

In this chapter we focus on one aspect of this problem that involves presenting an audio signal at selected directions in the room, while simultaneously minimizing the signal at other directions. For example, in home theater or television viewing applications a listener in a specific location in the room may not want to listen to the audio signal being transmitted, whereas another listener at a different location would prefer to listen to the signal. Consequently, if the objective is to keep one listener in a region with a reduced sound pressure level, then one can view this problem as that of signal cancellation in the direction of that listener. Similar applications arise in the automobile (e.g., when only the driver would prefer to listen to an audio signal), or any other environment with multiple listeners in which only a subset wish to listen to the audio signal.

Several methods have been proposed in the literature to lower the signal level either globally or in a local space within a region. Elliott and Nelson [105] proposed a global active power minimization technique for reducing the time-averaged acoustic pressure from a primary source in an enclosure, using a set of secondary source distributions. This least squares-based technique demonstrated that reduction in potential energy (and therefore sound pressure) can be achieved if the secondary sources are separated from the primary source by a distance which is less than half the wavelength of sound at the frequency of interest. It was suggested that this method can be employed to reduce the cockpit noise in a propeller-powered aircraft. Similarly, Ross [106] suggested the use of a filter that can minimize the signal power in the lobby of a building due to a generator outside the lobby by blocking the dominant plane wave mode with a loudspeaker. The reader is referred to several other interesting tutorial papers that have been published in active noise control [107, 108, 109]. Other examples could include head-mounted reference sensors using adaptive beamforming techniques [110].

In this chapter, the problem of signal cancellation is tackled by designing objective functions (criteria) that aim at reducing the sound pressure levels of signals in predetermined directions. A first objective criterion is designed for maximizing the difference in signal power between two different listener locations that have different source and receiver response characteristics. Thus, one application of this system lies in an environment having conflicting listening requirements, such as those mentioned earlier (e.g., automobiles, home environment). The filter, known as the eigenfilter, that is derived by optimizing the objective function, operates on the raw signal before being linearly transformed by the room responses in the direction of the listeners. Such filters aim at increasing the relative gain in signal power between the two listeners with some associated tradeoffs such as: (i) spectral distortion that may arise from the presence of the eigenfilter, and (ii) the sensitivity of the filter to the length of the room impulse response (reverberation). Further issues that can be researched, and which are beyond the scope of this chapter, include human perception of loudness, as well as perceptual aspects, such as coloration, and speech intelligibility.

The organization of this chapter is as follows. In the next section, we derive the required eigenfilter from a proposed objective function, and prove some of the theoretical properties of such filters. We provide experimental results for the performance (and tradeoff) of the eigenfilters in two situations: (i) using a synthesized room impulse response with a speech excitation, and (ii) using an actual room impulse response with a stochastic excitation. We also investigate the performance differences that are observed when using a minimum-phase model for the room response. We conclude the chapter by discussing some future research directions for selective signal cancellation using eigenfilters.

9.2 Traditional Methods for Acoustic Signal Cancellation

We divide the existing methods for acoustic signal cancellation as belonging to either (a) physical interfaces for acoustical signal cancellation, or (b) loudspeaker-based interfaces for acoustical signal cancellation in relatively large enclosures. Before introducing the different methods, we define two broad terms for signal cancellation.

Definition 1. An *active sound control* technique is a method for attenuating an unwanted acoustical signal (disturbance) by the introduction of controllable "secondary sources", whose outputs are arranged to interfere destructively with the disturbance.

Definition 2. A *passive sound control* technique is a method for attenuating an unwanted acoustical signal by the introduction of physical barriers of certain surface density [127].

All of the methods in this proposal belong to either or both of the aforementioned broad definitions for acoustic signal cancellation. Thus, typically the physical interfaces in (a) above are part of the passive sound control strategy; whereas, the loudspeaker-based interfaces are part of the active sound control strategy.

9.2.1 Passive Techniques

Simple Cotton Wad

This is a well-known and by far the cheapest passive sound control method for *absorbing* unwanted audio signals. In this method, an uninterested listener places a cotton ball inside each ear to limit the intensity (increase the attenuation) of the sound signal that enters the ear canal and strikes the ear drums. The disadvantage of this method is that the attenuation of the wad of cotton decreases with decrease in frequency. So, the wad of cotton is not quite good for passively canceling acoustic signals at low frequencies. Moreover, the insertion of the cotton ball can cause discomfort.

Ear Defenders

Ear defender [128] is a term used to designate a device that introduces attenuation of sound between a point outside the head and the eardrum. There are two types, namely, (i) the cushion type, and (ii) the insert type. The cushion type is similar to a pair of headphones with soft cushion ear pads. The cushion types are heavy and cumbersome. The insert type is a form of a plug that is pushed into the ear canal. Soft plastics and synthetic rubbers are the commonly used materials for the insert type defenders. A good ear defender will introduce an attenuation of 30 to 35 dB from 60 Hz to 8 kHz. However, the insertion may cause discomfort. Furthermore, they need to be custom made depending on the size of the ear.

Acoustic Barriers

Simply put, an acoustic barrier is a glorified wall. The aim of building a barrier is to redirect acoustic signal power generated by a source away from an uninterested listener. To be effective at this, the barrier must be constructed from "heavy" material (i.e., having a high surface density). Clearly, this is prohibitive in a room or an automobile. Moreover, a reasonable sized wall provides something of the order of 10 dB of attenuation in acoustic signal pressure levels. Attenuation of 20 dB or more is almost impossible to achieve with a simple barrier.

Sound Absorption Materials

Sound absorption can be achieved by introducing porosity in a material. Examples of simple sound absorbers are clothing material and an open window. Another example of a sound absorber (albeit ineffective) is a wall [11, p. 139]. The absorption of sound energy by the wall manifests as vibrational energy of the wall. This vibrational energy is then reradiated to the outside. The absorption α for a simple wall is given as

$$\alpha = \left(\frac{2c\rho_0}{M\omega}\right)^2 \tag{9.1}$$

where M denotes the mass per unit area of the wall, $\omega = 2\pi f$ is the angular frequency in rad/s, c is the speed of sound in air, and ρ_0 is the static density of air. Clearly, lower frequencies are better absorbed relative to high frequencies.

9.2.2 Active Techniques

Presence of Secondary Loudspeakers

There are several variants in this approach containing at least one secondary loudspeaker. Historically, the fundamental concept was first presented in a patent granted to Lueg [130], wherein Lueg suggested using a loudspeaker for canceling a one-dimensional acoustic wave, where a source generates a primary acoustic waveform $p(x, t) = p(x)e^{j\omega t}$,[1] expressed as an instantaneous sound pressure, in a duct (the solid line indicates the primary waveform). A microphone located farther downstream in the duct detects the acoustic signal. The output from the microphone is used to drive a loudspeaker, a secondary source, after being manipulated by a controller. The output from the loudspeaker is another acoustic signal $s(x, t) = s(x)e^{j\omega t}$ indicated by the dotted line. The loudspeaker is positioned and the controller is designed in a manner such that the secondary source $s(x, t)$ generates a signal that is of the same amplitude but opposite in phase as the primary source. That is,

---

[1] The decomposition of the acoustic wave into its time-dependent and frequency-dependent components has its origins in the solution to the one-dimensional wave equation expressed as $\partial^2 p(x, t)/\partial x^2 - \partial^2 p(x, t)/c_0^2 \partial t^2 = 0$, with c_0 being the speed of the acoustic wave in the given medium.

$$s(x) = -p(x) \tag{9.2}$$

The two acoustic signals are essentially designed to interfere destructively, which significantly attenuates the sound wave propagating downstream, relative to the secondary source, in the duct. However, what happens upstream, relative to the secondary source, is a completely different issue. It can be easily shown that the resulting magnitude of the pressure, upstream ($x < 0$) of the secondary loudspeaker (assuming a secondary loudspeaker located at $x = 0$) when $p(x) = Ae^{-jkx}$ ($k = 2\pi/\lambda$ is the wavenumber and λ is the wavelength), is $|r(x)| = |p(x) + s(x)| = 2A|\sin(kx)|$. Thus, the resulting absolute pressure, upstream, is twice as much when $x/\lambda = -0.5(n + 1/2)$, $n \in \{0, 1, 2, \ldots\}$. For this reason global control strategies are used to compensate for such effects when using secondary speakers, assuming one wishes to control tonal disturbances over a large region.

Active control strategies for three-dimensional wavefronts emphasize optimizing some objective criteria such as total power (sum of powers of primary and secondary wavefronts). The interested reader is referred to [129] for details of various optimization approaches, using multiple secondary speakers, in active sound control.

9.2.3 Parametric Loudspeaker Array

Recently new technologies in loudspeaker design have resulted in a parametric approach employing a grid of transducers that generate ultrasonic signals. These signals show up being very focussed in the loudspeaker directivity pattern. Sometimes the phrases "Audio Spotlight Devices" or "HyperSonic Sound Systems" are used when referring to them.[2] In essence, these loudspeakers are able to focus a narrow "beam" of sound in the direction of a specific listener.

9.3 Eigenfilter Design for Conflicting Listener Environments

9.3.1 Background

In this chapter, we primarily address the issue of designing eigenfilters for single source and dual listener environments (see Fig. 9.1). It is well established from linear system theory that

$$y_i(n) = \sum_{k=0}^{P-1} h_i(k)x(n-k) + v_i(n) \qquad i = 1, 2 \tag{9.3}$$

where $x(n)$ is the primary signal transmitted by a source, such as a loudspeaker; $y_i(n)$ is the signal received at listener R_i; h_i is the room transmission characteristic or room impulse response (modeled as a finite impulse response) between the source and listener R_i; and v_i is additive (ambient) noise at listener R_i. In a reverberant environment, due to multipath effects, the room responses vary with even small changes in the source and receiver locations [119, 63, 11], and in general $h_1(n) \neq h_2(n)$.

---

[2] The interested reader is directed to http://www.atcsd.com, of American Technology Corporation for a white paper on this technology.

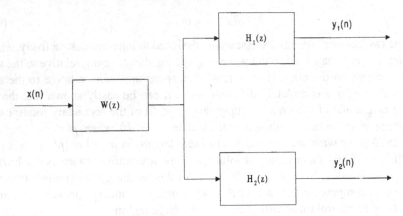

Fig. 9.1. The source and receiver model.

One method of modifying the transmitted primary signal $x(n)$ is to preprocess the source signal by a filter, called the eigenfilter, before transmitting it through the environment.

9.3.2 Determination of the Eigenfilter

Under our assumption of modeling the listeners as point receivers we can set up the problem as shown in Fig. 9.1, where $w_k; k = 0, 1, \ldots, M - 1$ represent the coefficients of the finite impulse response filter to be designed. For this problem, we assume that the receivers are stationary (i.e., the room impulse response for a certain (C, R) is time-invariant and linear, where C and R represent a source and a receiver), and the channel (room) impulse response is deterministic at the locations of the two listeners. The listening model is then simply related to (9.3), but the resulting transmitted primary signal is now filtered by w_k. Thus, the signal $y_i(n)$ at listener R_i, with the filter w_k present, is

$$y_i(n) = h_i(n) \otimes \sum_{k=0}^{M-1} w_k x(n - k) + v_i(n) \qquad i = 1, 2 \qquad (9.4)$$

where $\otimes$ represents the convolution operation. With this background, we view the signal cancellation problem as a gain maximization problem (between two arbitrary receivers); we can state the performance criterion as

$$J(n) = \max_{\underline{w}} \frac{1}{2} \left(\frac{\sigma_{y_2(n)}^2}{\sigma_{v_2(n)}^2} \right) - \frac{\lambda}{2} \left(\frac{\sigma_{y_1(n)}^2}{\sigma_{v_1(n)}^2} - \psi \right) \qquad (9.5)$$

in which we would like to maximize the signal-to-noise ratio (or signal power) in the direction of listener 2, while keeping the power towards listener 1 constrained at $10^{\psi_{dB}/10}$ (where $\psi_{dB} = 10 \log_{10} \psi$). In (9.5), $\sigma_{y_i(n)}^2 / \sigma_{v_i(n)}^2$ denotes the transmitted signal to ambient noise power at listener R_i with $y_i(n)$ as defined in (9.4). The quantity λ is the well-known Lagrange multiplier.

It is interesting to see that, when $x(n)$ and $v(n)$ are mutually uncorrelated, the two terms in the objective function (9.5) are structurally related to the mutual information between the source and listeners R_2 and R_1, respectively, under Gaussian noise assumptions [103].

Now observe that,

$$y_1(n) = h_1(n) \otimes \sum_{k=0}^{M-1} w_k x(n-k) + v_1(n) \tag{9.6}$$

where $h_1(n)$ is the room response in the direction for the listener labeled 1. Let $\underline{w} = (w_0, w_1, \ldots, w_{M-1})^T$, and $\underline{x}(n) = (x(n), x(n-1), \ldots, x(n-M+1))^T$; then (9.6) can be expressed as

$$
\begin{aligned}
y_1(n) &= h_1(n) \otimes \underline{w}^T \underline{x}(n) + v_1(n) \\
&= h_1(n) \otimes z(n) + v_1(n) \\
&= \sum_{p=0}^{L-1} h_1(p) z(n-p) + v_1(n)
\end{aligned}
\tag{9.7}
$$

where $z(n) = \underline{w}^T \underline{x}(n)$. We assume that the zero mean noise and signal are real and statistically independent (and uncorrelated in the Gaussian case). In this case signal power in the direction of listener 1 is

$$
\begin{aligned}
\sigma_{y_1(n)}^2 &= \mathrm{E}\left\{ \sum_{p=0}^{L-1} \sum_{q=0}^{L-1} h_1(p) h_1(q) z(n-p) z^T(n-q) \right\} + \sigma_{v_1(n)}^2 \\
&= \sum_{p=0}^{L-1} \sum_{q=0}^{L-1} h_1(p) h_1(q) (\underline{w}^T \mathbf{R}_{\underline{x}}(p,q) \underline{w}) + \sigma_{v_1(n)}^2
\end{aligned}
\tag{9.8}
$$

where $\underline{w} \in \Re^M$, $\mathbf{R}_{\underline{x}}(p,q) \in \Re^{M \times M}$, and

$$
\begin{aligned}
\mathbf{R}_{\underline{x}}(p,q) &= \mathrm{E}\{\underline{x}(n-p) \underline{x}^T(n-q)\} \\
\underline{x}(n-l) &= (x(n-l), \ldots, x(n-l-M+1))^T
\end{aligned}
\tag{9.9}
$$

Similarly,

$$\sigma_{y_2(n)}^2 = \sum_{p=0}^{S-1} \sum_{q=0}^{S-1} h_2(p) h_2(q) (\underline{w}^T \mathbf{R}_{\underline{x}}(p,q) \underline{w}) + \sigma_{v_2(n)}^2 \tag{9.10}$$

Solving $\nabla_{\underline{w}} J(n) = 0$, will provide the set of optimal tap coefficients. Hence from (9.5), (9.8), and (9.10), we obtain

$$\frac{\partial J(n)}{\partial \underline{w}} = \frac{1}{\sigma_{v_2(n)}^2} \sum_{p=0}^{S-1} \sum_{q=0}^{S-1} h_2(p) h_2(q) \mathbf{R}_{\underline{x}}(p,q) \underline{w}^*$$

$$-\frac{\lambda}{\sigma_{v_1(n)}^2} \sum_{p=0}^{L-1} \sum_{q=0}^{L-1} h_1(p)h_1(q)\mathbf{R}_{\underline{x}}(p,q)\underline{w}^* = 0 \qquad (9.11)$$

where $\underline{w}^*$ denotes the optimal coefficients. Let

$$A = \sum_{p=0}^{S-1} \sum_{q=0}^{S-1} h_2(p)h_2(q)\mathbf{R}_{\underline{x}}(p,q)$$

$$B = \sum_{p=0}^{L-1} \sum_{q=0}^{L-1} h_1(p)h_1(q)\mathbf{R}_{\underline{x}}(p,q) \qquad (9.12)$$

By assuming equal ambient noise powers at the two receivers (i.e., $\sigma_{v_2(n)}^2 = \sigma_{v_1(n)}^2$), (9.11) can be written as

$$\left.\frac{\partial J(n)}{\partial \underline{w}}\right|_{\underline{w}=\underline{w}^*} = (B^{-1}A - \lambda I)\underline{w}^* = 0 \qquad (9.13)$$

The reason for arranging the optimality condition in this fashion is to demonstrate that the maximization is in the form of an eigenvalue problem (i.e., the eigenvalues corresponding to the matrix $B^{-1}A$), with the eigenvectors being $\underline{w}^*$. There are in general M distinct eigenvalues for the $M \times M$ matrix, $B^{-1}A$, with the largest eigenvalue corresponding to the maximization of the ratio of the signal powers between receiver 2 and receiver 1. The optimal filter that yields this maximization is given by

$$\underline{w}^* = \underline{e}_{\lambda_{max}[B^{-1}A]} \qquad (9.14)$$

where $\underline{e}_{\lambda_{max}[B^{-1}A]}$ denotes the eigenvector corresponding to the maximum eigenvalue λ_{max} of $B^{-1}A$. An FIR filter whose impulse response corresponds to the elements of an eigenvector is called an *eigenfilter* [115, 8]. Finally, the gain between the two receiver locations can be expressed as

$$G_{dB} = 10\log_{10}\frac{\sigma_{y_2(n)}^2}{\sigma_{y_1(n)}^2} = 10\log_{10}\frac{\underline{w}^{*T}A\underline{w}^*}{\underline{w}^{*T}B\underline{w}^*} \qquad (9.15)$$

Clearly it can be seen from (9.14) that the optimal filter coefficients are determined by the channel responses between the source and the two listeners. The degrees of freedom for the eigenfilter are the order M of the eigenfilter.

Fundamentally, by recasting the signal cancellation problem as a gain maximization problem, we aim at introducing a gain of G dB between two listeners, R_1 and R_2. This G dB gain is equivalent to virtually positioning listener R_1 at a distance that is $\sqrt{10^{G_{dB}/10}}$ the distance of listener R_2 from a fixed sound source C.[3] This is depicted in Fig. 9.2, where R1 (solid head) is experiencing signal power levels that he would expect if he were positioned at $\sqrt{10^{G_{dB}/10}}$ (indicated by the dotted head).

---

[3] Strictly speaking, in the free field, the gain based on the inverse square law is expressed as $Q = 10\log_{10} r_1^2/r_2^2$ (dB), where r_1, r_2 are the radial distances of listeners R_1 and R_2 from the source.

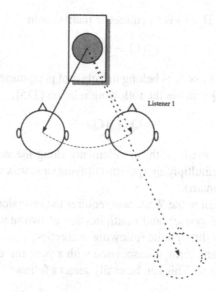

Fig. 9.2. The effect of gain maximization.

9.3.3 Theoretical Properties of Eigenfilters

Some interesting properties of the proposed eigenfilter emerge under wide-sense stationary (WSS) assumptions. In this section we derive some properties of eigenfilters for selective signal cancellation, which we then use in a later section.

In signal processing applications, the statistics (ensemble averages) of a stochastic process are often independent of time. For example, quantization noise exhibits constant mean and variance, whenever the input signal is "sufficiently complex." Moreover, it is also assumed that the first-order and second-order probability density functions (PDFs) of quantization noise are independent of time. These conditions impose the constraint of stationarity. Because we are primarily concerned with signal power, which is characterized by the first-order and second-order moments (i.e., mean and correlation), and not directly with the PDFs, we focus on the wide-sense stationarity aspect. It should be noted that in the case of Gaussian processes, wide-sense stationarity is equivalent to strict-sense stationarity, which is a consequence of the fact that Gaussian processes are completely characterized by the mean and variance. Below, we provide some definitions, properties, and a basic theorem pertaining to eigenfilter structure for WSS processes.

Property 1 : For a WSS process $x(n)$, and $y(n)$ with finite variances, the matrix $\mathbf{R}_{\underline{x}}(p, q)$ is Toeplitz, and the gain (9.15) can be expressed as

$$G_{dB} = 10 \log_{10} \frac{\int_{2\pi} |W^*(e^{j\omega})|^2 |H_2(e^{j\omega})|^2 S_x(e^{j\omega}) d\omega}{\int_{2\pi} |W^*(e^{j\omega})|^2 |H_1(e^{j\omega})|^2 S_x(e^{j\omega}) d\omega} \qquad (9.16)$$

where $r_x(k) \in \mathbf{R}_{\underline{x}}(k)$ and $S_x(e^{j\omega})$ form a Fourier transform pair, and $h_1(n)$ and $h_2(n)$ are stable responses. Moreover, because we are focusing on real processes in this chapter, the matrix $\mathbf{R}_{\underline{x}}(k)$ is a symmetric matrix, with

$$r_x(k) = r_x(-k) \tag{9.17}$$

Property 2 : Toeplitz matrices belong to a class of persymmetric matrices. A $p \times p$ persymmetric matrix $\mathbf{Q}$ satisfies the following relation [115],

$$\mathbf{Q} = \mathbf{JQJ} \tag{9.18}$$

where $\mathbf{J}$ is a diagonal matrix with unit elements along the northeast-to-southwest diagonal. Basically, premultiplying (postmultiplying) a matrix with $\mathbf{J}$ exchanges the rows (columns) of the matrix.

The eigenfilter design in the WSS case requires the inversion of a scaled Toeplitz matrix (via the room response), and multiplication of two matrices. We investigate these operations briefly through the following properties.

Property 3 : A scaling term, c, associated with a persymmetric matrix leaves its persymmetricity unaltered. This can be easily seen as follows,

$$\mathbf{J}c\mathbf{QJ} = c\mathbf{JQJ} = c\mathbf{Q} \tag{9.19}$$

Property 4 : Linear combination of persymmetric matrices yields a persymmetric matrix.

$$\mathbf{J}[c_1\mathbf{Q}_1 + c_2\mathbf{Q}_2]\mathbf{J} = c_1\mathbf{JQ}_1\mathbf{J} + c_2\mathbf{JQ}_2\mathbf{J} = c_1\mathbf{Q}_1 + c_2\mathbf{Q}_2 \tag{9.20}$$

Hence, from the above properties, the matrices A and B (in (9.12)) are persymmetric.

Property 5 : The inverse of a persymmetric matrix is persymmetric.

$$\mathbf{Q} = \mathbf{JQJ}$$
$$\mathbf{Q}^{-1} = (\mathbf{JQJ})^{-1} = \mathbf{J}^{-1}\mathbf{Q}^{-1}\mathbf{J}^{-1} = \mathbf{JQ}^{-1}\mathbf{J} \tag{9.21}$$

Property 6 : The product of persymmetric matrices is persymmetric.

$$\mathbf{Q}_1\mathbf{Q}_2 = \mathbf{JQ}_1\mathbf{JJQ}_2\mathbf{J} = \mathbf{JQ}_1\mathbf{Q}_2\mathbf{J}$$

where we have used the fact that $\mathbf{JJ} = \mathbf{J}^2 = I$. Thus, $B^{-1}A$ is persymmetric.

Theorem 9.1. *The roots of the eigenfilter corresponding to a distinct maximum eigenvalue, lie on the unit circle for a Toeplitz* $\mathbf{R}_{\underline{x}}(p,q) = \mathbf{R}_{\underline{x}}(k)$.

Proof : Because the matrix $B^{-1}A$ is persymmetric, based on properties 2 to 5, we can incorporate a proof similar to the one given in [113] as proof.

Property 7 [112] : If $\mathbf{Q}$ is persymmetric with distinct eigenvalues, then $\mathbf{Q}$ has $\lceil p/2 \rceil$ symmetric eigenvectors, and $\lfloor p/2 \rfloor$ skew symmetric eigenvectors, where $\lceil x \rceil$ ($\lfloor x \rfloor$) indicates the smallest (largest) integer greater (less) than or equal to x.

A persymmetric matrix is not symmetric about the main diagonal, hence the eigenvectors are not mutually orthogonal. However, in light of the present theory we can prove the following theorem.

Theorem 9.2. *Skew-symmetric and symmetric eigenvectors for persymmetric matrices are orthogonal to each other.*

Proof: Let

$$V_1 = \{\underline{w} : \mathbf{J}\underline{w} = \underline{w}\}$$
$$V_2 = \{\underline{w} : \mathbf{J}\underline{w} = -\underline{w}\} \tag{9.22}$$

Now,

$$\mathbf{J}\underline{\nu}_1 = \underline{\nu}_1 \quad \underline{\nu}_1 \in V_1 \tag{9.23}$$

then with $\underline{\nu}_2 \in V_2$ we have,

$$\underline{\nu}_2^T \mathbf{J}\underline{\nu}_1 = \underline{\nu}_2^T \underline{\nu}_1 \tag{9.24}$$

But,

$$\mathbf{J}\underline{\nu}_2 = -\underline{\nu}_2 \Rightarrow \underline{\nu}_2^T \mathbf{J} = -\underline{\nu}_2^T \tag{9.25}$$

using the fact the $\mathbf{J}^T = \mathbf{J}$. Substituting (9.25) into (9.24) results in

$$-\underline{\nu}_2^T \underline{\nu}_1 = \underline{\nu}_2^T \underline{\nu}_1 \Rightarrow \underline{\nu}_2^T \underline{\nu}_1 = 0 \tag{9.26}$$

which proves the theorem.

Property 8 : From the unit norm property of eigenfilters ($\|\underline{w}^*\|^2 = 1$), and *Parsevals* relation, we have

$$\int_{2\pi} |W^*(e^{j\omega})|^2 d\omega = 2\pi \tag{9.27}$$

Property 9 [112] : The eigenvectors associated with $B^{-1}A$ satisfy either,

$$\mathbf{J}\underline{w} = \begin{cases} \underline{w} & \text{symmetric} \\ -\underline{w} & \text{skew-symmetric.} \end{cases} \tag{9.28}$$

Theorem 9.3. *The optimal eigenfilter (9.14) is a linear phase FIR filter having a constant phase and group delay (symmetric case), or a constant group delay (skew-symmetric case).*

Proof:

$$w^*(m) = \begin{cases} w^*(M-1-m) & \text{symmetric} \\ -w^*(M-1-m) & \text{skew-symmetric} \end{cases} \quad m = 0, 1, \dots, M-1 \tag{9.29}$$

because $\mathbf{J}$, in property 9, exchanges the elements of the optimal eigenfilter.

In the following section we discuss the results of the designed eigenfilter for a speech source.

9.4 Results

The degrees of freedom for the eigenfilter in (9.14), is the order M. Variabilities such as (i) the choice for the modeled duration (S, L) for the room responses (9.12), (ii) the choice of the impulse response (i.e., whether it is minimum-phase or nonminimum-phase), and (iii) variations in the room response due to listener (or head) position changes affect the performance (gain). We study (i) and (ii) in the present chapter with the assumption of $L = S$. The choice for the filter order and the modeled impulse response duration affects the gain (9.15) and distortion (defined later in this section) of the signal at the microphones. Basically, a lower duration response used for designing the eigenfilter will reduce the operations for computing the eigenfilter, but may affect performance. In summary, the length of the room response (reverberation) modeled in the design of the eigenfilter affects the performance and this variation in performance is referred to as the sensitivity of the eigenfilter to the length of the room response.

9.4.1 Eigenfilter Performance as a Function of Filter Order M

In this experiment, the excitation, $x(n)$, was a segment of speech signal obtained from [120]. The speech was an unvoiced fricated sound /S/ as in "sat" obtained from a male subject and is shown in Fig. 9.3.

As is well known, this sound is obtained by exciting a locally time-invariant, causal, stable vocal tract filter by a stationary uncorrelated white noise sequence, which is independent from the vocal tract filter [114]. The stability of the vocal tract

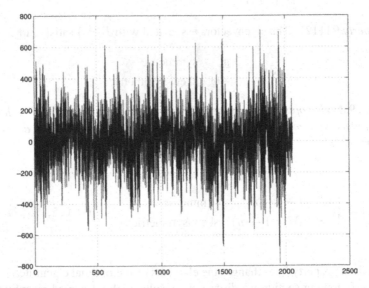

Fig. 9.3. The speech signal segment for the unvoiced fricative /S/ as in *sat*.

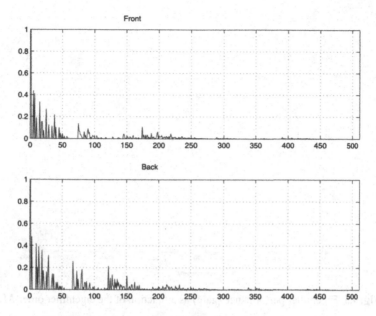

Fig. 9.4. Impulse responses for the front and back positions.

filter is essential, as it guarantees the stationarity of the sequence $x(n)$ [122]. The impulse responses were generated synthetically from the room acoustics simulator software [123]. The estimation of these responses was based on the image method (geometric modeling) of reflections created by ideal omnidirectional sources, and received by ideal omnidirectional receivers [61]. For the present scenario the modeled room was of dimensions, 15 m × 10 m × 4 m. The source speaker was at (1 m, 1 m, 1 m) from a reference northwest corner. The impulse response for the "front" microphone located at (4.9 m, 1.7 m, 1 m) relative to the reference, was denoted as $h_2(n)$, and the "back microphone" located at (4.5 m, 6.4 m, 1 m) had impulse response measurement $h_1(n)$. The two responses are plotted as positive pressure amplitudes in Fig. 9.4 (ignoring the initial delay). This situation is similar to the case for listeners in an automobile, where the front left speaker is active, and the relative gain to be maximized is between the front driver and the back passenger.

A plot of the gain (9.15) as a function of the filter order for the aforementioned signal and impulse responses is shown in Fig. 9.5.

Firstly, a different microphone positioning will require a new simulation for computing (9.14), and determining the performance thereof. Secondly, larger duration filters increase the gain, but affect the signal characteristics at the receiver in the form of distortion. Basically, a distortion measure is an assignment of a nonnegative number between two quantities to assess their fidelity. According to Gray *et al.* [124], a distortion measure should satisfy the following properties: (1) it must be meaningful, in that small and large distortions between the two quantities correspond to good and bad subjective quality, (2) it must be tractable and should be easily tested via mathematical analysis, and (3) it must be computable (the actual distortions in a real system

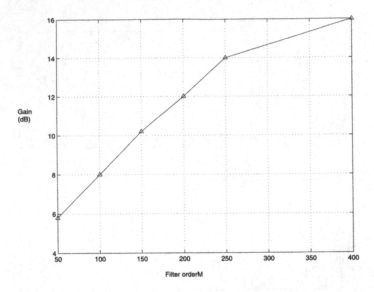

Fig. 9.5. Eigenfilter performance (gain) as a function of the eigenfilter order M.

can be efficiently computed). The proposed distortion measure is evaluated in terms of an $L_p, (p = 1)$ norm on $(-\pi, \pi)$ [125] and models the variation in the received spectrum at listener 2,[1] due to the presence of the eigenfilter, over the natural event, that of the absence of the filter. We use the L_1 norm due to its ease of analysis and computation for the current problem. Before presenting the results for the distortion against filter order, we prove that the average spectrum error (stated in terms of the spectral local matching property [121]) E_M is constant for any eigenfilter order.

Theorem 9.4. *The spectrum error E_M defined in terms of the spectral match is*

$$E_M = \left\| \frac{S_{\hat{y}}(e^{j\omega})}{S_y(e^{j\omega})} \right\|_1 = 1, \qquad \forall \, M \tag{9.30}$$

for an Mth order eigenfilter, and

$$S_{\hat{y}}(e^{j\omega}) = |H_2(e^{j\omega})|^2 |W_M(e^{j\omega})|^2 S_x(e^{j\omega}) = |W_M(e^{j\omega})|^2 S_y(e^{j\omega}) \tag{9.31}$$

where $S_{\hat{y}}(e^{j\omega})$, $S_y(e^{j\omega})$ are the spectra associated with the presence and absence of the eigenfilter, respectively (an equivalent model is shown in Fig. 9.6), and $W_M(e^{j\omega}) = \sum_{i=0}^{M-1} w_i e^{-j\omega i}$.

Proof: From the L_1 definition, we have

$$E_M = \int_{-\pi}^{\pi} \left| \frac{S_{\hat{y}}(e^{j\omega})}{S_y(e^{j\omega})} \right| \frac{d\omega}{2\pi} \tag{9.32}$$

---

[1] The evaluation of the distortion at listener 1 is not important, because the intention is to "cancel" the signal in her direction.

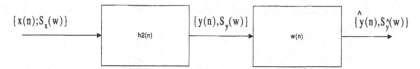

Fig. 9.6. Equivalent spectral model in the direction of listener 2 using the eigenfilter w_k.

From (9.28), (9.31), and (9.32) it can be seen that

$$E_M = \int_{-\pi}^{\pi} |W_M(e^{j\omega})|^2 \frac{d\omega}{2\pi} = 1 \qquad (9.33)$$

It is interesting to observe that a similar result can be established for the linear prediction spectral matching problem [121]. Also, when the FIR eigenfilter is of the lowest order with $M = 1$, and $w_0 = 1$, then the impulse response of the eigenfilter is $w(n) = \delta(n)$, and E_1 is unity (observe that with $w(n) = \delta(n)$ we have $h_2(n) \otimes \delta(n) = h_2(n)$).

An interpretation of (9.33) is that irrespective of the filter order ($M > 1$), the average spectral ratio is unity, which means that in terms of the two spectra, $S_{\hat{y}}(e^{j\omega})$ will be greater than $S_y(e^{j\omega})$ in some regions, and less in other regions, such that (9.33) holds.

The *log-spectral distortion* $d_M(S_{\hat{y}}(e^{j\omega}), S_y(e^{j\omega}))$ for an eigenfilter of order M on an L_1 space is defined as

$$
\begin{aligned}
d_M(S_{\hat{y}}(e^{j\omega}), S_y(e^{j\omega})) &= \left\| \log S_y(e^{j\omega}) - \log S_{\hat{y}}(e^{j\omega}) \right\|_1 \\
&= \left\| \log S_{\hat{y}}(e^{j\omega})/S_y(e^{j\omega}) \right\|_1 \\
&= \left\| \log |W_M(e^{j\omega})|^2 \right\|_1 \\
&= \int_{-\pi}^{\pi} \left| \log |W_M(e^{j\omega})|^2 \right| \frac{d\omega}{2\pi} \qquad (9.34)
\end{aligned}
$$

It can be easily shown that $d_M(S_{\hat{y}}(e^{j\omega}), S_y(e^{j\omega})) \geq 0$, with equality achieved when the eigenfilter is of unit order with $w_0 = 1$. In Fig. 9.7, we have computed the distortion (9.34), using standard numerical integration algorithms, as a function of the filter order for the present problem. Figure 9.8 summarizes the results from Fig. 9.9 and Fig. 9.7, through the gain-distortion *constellation* diagram. Thus depending on whether a certain amount of distortion is allowable, we can choose a certain point in the constellation (distortionless performance is obtained for the point located along the positive ordinate axis in the constellation).

Clearly, there is an improvement in the gain-to-distortion ratio with the increase in filter order (e.g., from Fig. 9.8, $M = 400$ gives a gain-to-distortion ratio of $10^{1.6}/9.8 \approx 4$, whereas $M = 250$ gives the gain-to-distortion ratio as 3). Also, for example, with filter order $M = 400$, the relative gain between the two locations is as much as 16 dB. This ideally corresponds to a virtual position of listener 1, for whom the sound cancellation is relevant, to be at a distance that is *four* times as far from a fixed source as the other listener (listener 2).

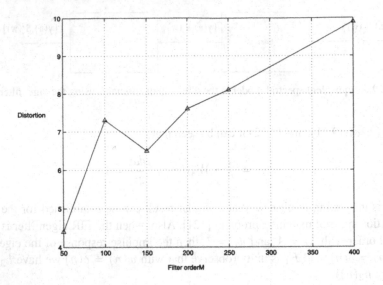

Fig. 9.7. Eigenfilter distortion as a function of the eigenfilter order M.

9.4.2 Performance Sensitivity as a Function of the Room Response Duration

From Eqs. (9.12), (9.14), and (9.15) we see that the eigenfilter performance can be affected by (i) the room response duration modeled in the eigenfilter design, as well as (ii) the nature of the room response (i.e., whether it is characterized by an equiv-

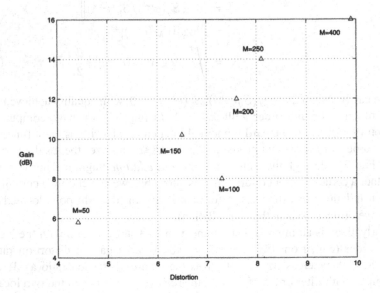

Fig. 9.8. Gain-to-distortion constellation space. Distortionless performance is obtained along the positive ordinate axis.

alent minimum phase model). In summary, a short duration room response if used in (9.12), for determining (9.14), will reduce the computational requirements for designing the eigenfilter. However, this could reduce the performance because the eigenfilter does not use all the information contained in the room responses. This then introduces a performance tradeoff. The question then is, can an eigenfilter (9.14) be designed with short duration room

response (for savings in computation) in the A and B matrices in (9.12), but yet does not cause the performance (9.15) to be affected. Of course, care should be taken to evaluate the performance in that the A and B matrices in (9.15) should have the full duration room responses.

To understand this performance tradeoff, we design the eigenfilter of length $M < L$ (L being the actual duration of the room impulse responses in the two directions), based on windowing both room responses with the window being rectangular and having duration $P < L$. We then analyze the performance (9.15) of the filter to increasing room response length. Basically the goal of this experiment is, can we design an eigenfilter with sufficiently short room responses (in (9.14)) without compromising the performance? To answer this question, the following procedure is adopted.

(a) Design the eigenfilter $\underline{\hat{w}}^* \in \Re^{M \times 1}$ for a shortened room response duration $P < L$,

$$\underline{\hat{w}}^* = \underline{e}_{\lambda_{max}[\hat{B}^{-1}\hat{A}]} \tag{9.35}$$

with

$$\hat{A} = \sum_{p=0}^{P-1} \sum_{q=0}^{P-1} h_2(p)h_2(q)\mathbf{R}_{\underline{x}}(p,q)$$

$$\hat{B} = \sum_{p=0}^{P-1} \sum_{q=0}^{P-1} h_1(p)h_1(q)\mathbf{R}_{\underline{x}}(p,q) \qquad M \le P < L \tag{9.36}$$

where the hat above the matrices in (9.36) denotes an approximation to the true quantities in (9.12), and the corresponding eigenfilter (9.35) is the resulting approximation (due to reduced duration $P < L$) to (9.14). We have included the constraint $M \le P < L$ to keep the order of the eigenfilter low (reduced processing), for a given real room response duration $L = 8192$, as explained below.

(b) Evaluate the performance (9.15) of the filter with the true matrices A and B (9.12) containing the full duration room responses.

We consider the performance when we select the responses according to (a) $h_i(n) = h_{i,min}(n) \otimes h_{i,ap}(n)$, and b) $h_i(n) = h_{i,min}(n); i = 1, 2$; where $h_{i,min}(n)$ and $h_{i,ap}(n)$ are the minimum-phase and all-pass components of the room responses. The impulse responses $h_1(n)$ and $h_2(n)$ (comprising 8192 points) were obtained in a highly reverberant room from the same microphones.

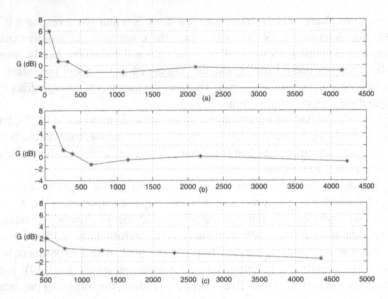

Fig. 9.9. $M = 64$; (a) $P = 64$; (b)$P = 128$; (c)$P = 512$.

Impulse Response $h_i(n) = h_{i,min}(n) \otimes h_{i,ap}(n); i = 1, 2$

In Fig. 9.9, we show the performance of the eigenfilter design as a function of the length of the impulse response. The length of the FIR filter was $M = 64$. The performance in each subplot as a function of the impulse response increments is shown, where we chose $\Delta P = \{0\} \cup \{2^k : k \in [7, 12], k \in I\}$, where I denotes the integer set. Thus, Fig. 9.9(a) represents an eigenfilter of length $M = 64$ designed with duration P, of the windowed impulse response, to be 64 (after removing the pure delay). The second performance evaluation, marked by an asterisk, is at $P + \Delta P = 64 + 2^7 = 192$. In Fig. 9.10 and Fig. 9.11, we show the sensitivity of the eigenfilter for filter lengths $M = 128$ and $M = 256$ for various windowed room impulse responses.

From the figures, we confirmed a better gain performance with increased filter length. By considering a larger duration room impulse response in the eigenfilter design, we lower the gain relatively but improve its evenness (flatness), Ideally, we want a small duration filter length (relative to the length of the room responses) with a large gain and uniform performance (low sensitivity to the length of the room impulse response).

Impulse Response $h_i(n) = h_{i,min}(n); i = 1, 2$

In Figs. 9.12 to 9.14, we show the performance of the eigenfilter for various windowed room responses and with different filter lengths. The performance (in terms of uniformity and level of the gain) is better than the nonminimum-phase impulse response model. We need to investigate this difference in the future.

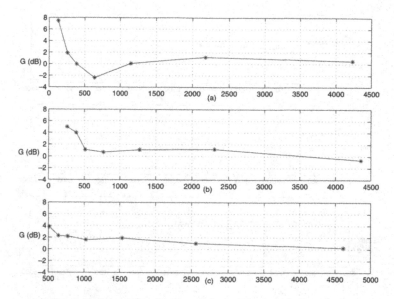

Fig. 9.10. $M = 128$; (a) $P = 128$; (b)$P = 256$; (c)$P = 512$.

9.5 Summary

There is a proliferation of integrated media systems that combine multiple audio and video signals to achieve tele-immersion among distant participants. One of the key aspects that must be addressed is the delivery of the appropriate sound to each local

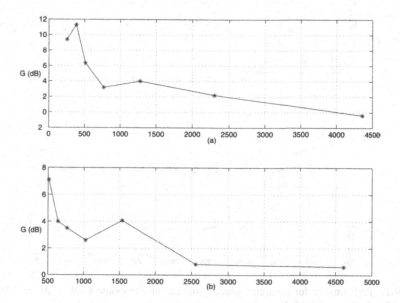

Fig. 9.11. $M = 256$; (a) $P = 256$; (b) $P = 512$.

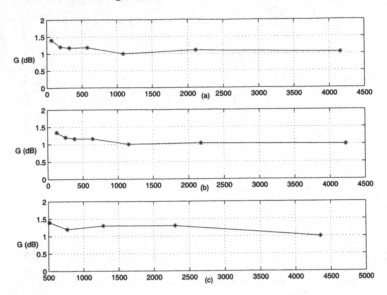

Fig. 9.12. Performance for minimum-phase room impulse response models. $M = 64$; (a) $P = 64$; (b) $P = 128$; (c) $P = 512$.

participant in the room. In addition, sound intended for other participants or originating from noise sources must be canceled. In this chapter we presented a technique for canceling audio signals using a novel approach based on information theory. We ad-

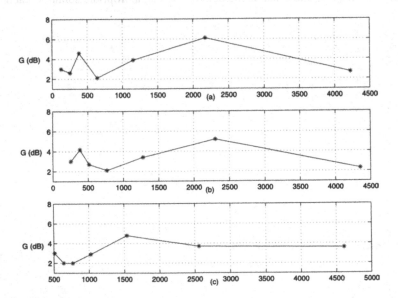

Fig. 9.13. Performance for minimum-phase room impulse response models. $M = 128$; (a) $P = 128$; (b) $P = 256$; (c) $P = 512$.

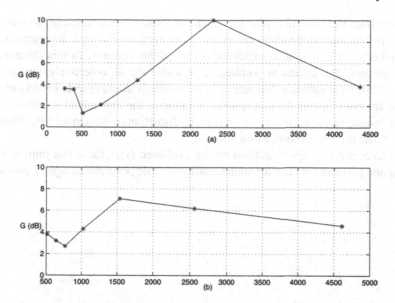

Fig. 9.14. Performance for minimum-phase room impulse response models. $M = 256$; (a) $P = 256$; (b) $P = 512$.

dressed this technique as the eigenfilter method, because the filter was derived based on maximizing the relative power between the two listeners in an acoustic enclosure. We also derived some of its theoretical properties (e.g., linear phase). For fixed room responses, we investigated (i) performance (gain) tradeoff to distortion, (ii) sensitivity of the performance to modeled room impulse duration. Our findings, for the present channel conditions, indicate that increasing the filter order improves the gain-to-distortion ratio. Thus, depending on the application, a suitable order filter may be chosen from the gain-to-distortion constellation diagram or from the sensitivity results. Furthermore, our findings for a particular scenario indicate that by extracting the minimum-phase component we get a better performance (in terms of uniformity and level of the gain) than the nonminimum-phase impulse response model.

In summary, this chapter addressed a fairly new application area, and clearly not all answers are contained. Hence, future directions include research in the following areas.

(a) The distortion measure that is introduced in Eq. (9.34) is easy to compute and is well known in the literature. Of course speech intelligibility is affected by a change in the frequency spectrum (this change in the spectrum is computed in the form of the distortion measure), and large changes will result in a degradation in speech intelligibility. To determine how large is large, as a next step one could perform speech intelligibility tests for consonants, for example, using a "confusion matrix".

(b) Investigation of the characteristics of *gain zones*, regions in space around the microphones which have a gain improvement of at least 10 dB in SPL, and viewing them from the acoustical physics viewpoint. Also the evaluation of the loudness (which is frequency-dependent) criteria using eigenfilters is a topic for research.

(c) Performing psychoacoustical/subjective measurements. In this chapter, we have addressed the effects of prefiltering an audio signal, objectively, through the spectral distortion measure. Subjective (double-blind) listening tests need to be performed for investigating the perceptual coloration of the transmitted signals.

(e) Investigation of the effects on gain-to-distortion by designing LPC filters to approximate the room transfer functions.

(f) Alternate objective functions can be evaluated (viz., those that minimize the SPL at one position but keep the sound quality at other positions as high as possible).

References

1. Mitra S (2001), Digital Signal Processing: A Computer Based Approach. McGraw-Hill.
2. Oppenheim A, Schafer R (1989), Discrete Time Signal Processing. Prentice-Hall.
3. Porat B (1996), A Course in Digitial Signal Processing, John Wiley & Sons.
4. Churchill R, Brown J (1989), Complex Variables and Applications, McGraw-Hill.
5. Rabiner L, Crochiere R (1983), Multirate Digital Signal Processing, Prentice-Hall.
6. Vaidyanathan PP (1993), Multirate Systems and Filter Banks, Prentice-Hall.
7. Mitra S, Kaiser JF (1993), Handbook for Digital Signal Processing, John Wiley & Sons.
8. Haykin S (1996), Adaptive Filter Theory, Prentice-Hall.
9. Hayes MH (1996), Statistical Digital Signal Processing and Modeling, John Wiley & Sons.
10. Rabiner L, Juang B-H (1993), Fundamentals of speech recognition, Prentice-Hall.
11. Kuttruff H (1991), Room Acoustics, Elsevier Applied Science.
12. Morse PM, Ingard KU (1986), Theoretical Acoustics, Princeton Univ. Press.
13. Widrow B., Hoff ME Jr. (1960), IRE WESCON Conv. Rec., Part 4:96–104.
14. Schroeder MR (1979), J. Acoust. Soc. Amer., 66:497–500.
15. Cook RK, Waterhouse RV, Berendt RD, Edelman S, and Thompson MC (1955), J. Acoust. Soc. Amer., 27(6):1072–1077.
16. Schroeder MR (1962), J. Acoust. Soc. Amer., 34(12):1819–1823.
17. Schroeder MR (1975), J. Acoust. Soc. Amer., 57:149–150.
18. Müller S, Massarani P (2001), J. Audio Eng. Soc., 49(6):443–471.
19. Dunn C, Hawksford MO (1993), J. Audio Eng. Soc., 41:314–335.
20. Farina A (Apr. 2000), 108th Conv. of Audio Eng. Soc. (preprint 5093).
21. Stan G-B, Embrechts J-J, and Archambeau D (2002), J. Audio Eng. Soc., 50(4):249–262.
22. Fletcher H, Munson WA (1933), J. Acoust. Soc. Amer., 5:82–108.
23. Robsinson DW, Dadson RS (1956), J. Appl. Phys., 7:166–181.
24. Intl. Org. for Standardization (1987), ISO-226.
25. Stevens SS (1972), J. Acoust. Soc. Amer., 51:575–601.
26. Moore BCJ (2000), An Introduction to the Psychology of Hearing, Academic Press.
27. Stephens SDG (1973), J. Sound Vib., 37:235–246.
28. Fletcher H (1940), Rev. Mod. Phys., 12:47–65.
29. Patterson RD, Moore BCJ (1986), in Frequency Selectivity in Hearing (Ed. Moore BCJ), Academic Press.
30. Hamilton PM (1957), J. Acoust. Soc. Amer., 29:506–511.
31. Greenwood DD (1961), J. Acoust. Soc. Amer., 33:484–501.

32. Patterson RD (1976), J. Acoust. Soc. Amer., 59:640–654.
33. Gierlich HW (1992), Appl. Acoust., 36:219-243.
34. Wightman FL, Kistler DJ (1989), J. Acoust. Soc. Amer., 85(2):858–867.
35. Wightman FL, Kistler DJ (1989), J. Acoust. Soc. Amer., 85(2):868-878.
36. Bauer BB (1961), J. Audio Eng. Soc., 9(2):148–151.
37. Moller H (1992), Appl. Acoust., 36:171–218.
38. Asano F, Suzuki Y, and Sone T (1990), J. Acoust. Soc. Amer., 88(1):159–168.
39. Middlebrooks JC, Green DM (1991), Annu. Rev. Psychol., 42:135–159.
40. Hebrank J, Wright D (1974), J. Acoust. Soc. Amer., 56(6):1829-1834.
41. Schroeder MR, Atal BS (1963), IEEE Conv. Record, 7:150–155, 1963.
42. Schroeder MR, Gottlob D, and Siebrasse KF (1974), J. Acoust. Soc. Amer., 56:1195–1201.
43. Cooper DH, Bauck J (1989), J. Audio Eng. Soc., 37(1/2):3–19.
44. Nelson PA, Hamada H, and Elliott SJ (1992), IEEE Trans. Signal Process., 40:1621–1632.
45. Lim J-S, Kyriakakis C (2000), 109th Conv. of Audio Eng. Soc. (preprint 5183).
46. Cooper DH, Bauck J (1996), J. Audio Eng. Soc., 44:683–705.
47. Itakura F, Saito S (1970), Elect. and Comm. in Japan, 53A:36–43.
48. Laroche J, Meillier JL (1994), IEEE Trans. on Speech and Audio Proc., 2.
49. Pozidis H, Petropulu AP (1997), IEEE Trans. on Sig. Proc., 45:2977–2993.
50. Moulines E, Duhamel P, Cardoso JF, and Mayrargue S (1995), IEEE Trans. on Sig. Proc., 43:516–525.
51. Widrow B, Walach E (1995), Adaptive Inverse Control, Prentice-Hall.
52. Mouchtaris A, Lim J-S, Holman T, and Kyriakakis C (1998), Proc. IEEE Mult. Sig. Proc. Wkshp. (MMSP '98).
53. Bershad NJ, Feintuch PL (1986), IEEE Trans. Acoust., Speech Sig. Proc., ASSP-34:452–461.
54. Ferrara ER Jr. (1980), IEEE Trans. Acoust., Speech Sig. Proc., ASSP-28(4):474–475.
55. Bershad NJ, Feintuch PL (1986), IEEE Trans. Acoust., Speech Sig. Proc., ASSP-34:452–461.
56. Widrow B, McCool JM (1976), IEEE Trans. Antennas and Propagation, AP-24(5):615–637.
57. Narayan SS, Peterson AM, and Narasimha MJ (1983), IEEE Trans. Acoust. Speech. Sig. Proc., ASSP-31(3):609–615.
58. Horowitz LL, and Senne KD(1981), IEEE Trans. Circuits and Syst., CAS-28(6):562–576, 1981.
59. Narayan SS, Peterson AM, and Narasimha MJ (1983), IEEE Trans. Acoust. Speech. Sig. Proc., ASSP-31(3):609–615.
60. Bharitkar S, Kyriakakis C (2003), Proc. 37th IEEE Asilomar Conf. on Sig. Syst. Comp., 1:546–549.
61. Allen JB, Berkley DA (1979), J. Acoust. Soc. Amer., 65:943–950.
62. Weiss S, Rice G, and Stewart RW (1999), IEEE Wkshp. Appl. Sig. Proc. Audio and Acoust., 203–206.
63. Mourjopoulos J (1985), J. Sound & Vib., 102(2):217–228.
64. Elliott SJ, Nelson PA (1989), J. Audio Eng. Soc., 37(11):899–907.
65. Mourjopoulos J (1994), J. Sound & Vib., 43(11).
66. Haneda Y, Makino S, and Kaneda Y (1994), IEEE Trans. on Speech and Audio Proc., 2(2):320–328.
67. Miyoshi M, Kaneda Y (1988), IEEE Trans. Acoust. Speech and Signal Proc., 36(2):145–152.

68. Haneda Y, Makino S, and Kaneda Y (1997), IEEE Trans. on Speech and Audio Proc., 5(4):325–333.
69. Neely S, Allen J (1979), J. Acoust. Soc. Amer., 66(1):165–169.
70. Radlović B, Kennedy R (2000), IEEE Trans. on Speech and Audio Proc., 8(6):728–737.
71. Karjalainen M, Piirilä E, Järvinen A, and Huopaniemi J (1999),J. Audio Eng. Soc., 47 (1/2):15–31.
72. Karjalainen M, Härmä A, Laine UK, and Huopaniemi J (1997), Proc. 1997 IEEE Wkshp. on Appl. Signal Proc. Audio and Acoust. (WASPAA '97).
73. Härmä A, Karjalainen M, Savioja L, Välimäki V, Laine UK, and Huopaniemi J (2000), J. Audio Eng. Soc., 48(11):1011–1031.
74. Chang PR, Lin CG, and Yeh BF (1994), J. Acoust. Soc. Amer., 95(6):3400–3408.
75. Mourjopoulos J, Clarkson P, and Hammond J (1982), Proc. ICASSP, 1858–1861.
76. Bezdek J (1981), Pattern recognition with fuzzy objective function algorithms, Plenum.
77. Dunn JC (1973), J. Cybern., 3:32–57.
78. Xie XL, Beni G (1991), IEEE Trans. on Pattern Analysis and Mach. Intelligence, 3:841–846.
79. Pal NR, Bezdek JC (1995), IEEE Trans. on Fuzzy Syst., 3(3):370–379.
80. Markel JD, Gray, AH Jr. (1976), Linear Prediction of Speech, Springer-Verlag.
81. Alku P, Bäckström T (2004), IEEE Trans. on Speech and Audio Proc., 12(2):93–99.
82. Oppenheim A, Johnson D, and Steiglitz K (1971), Proc. IEEE, 59:299–301.
83. Smith JO, Abel JS (1999), IEEE Trans. on Speech and Audio Proc., 7(6):697–708.
84. Zwicker E, Fastl H (1990), Psychoacoustics: Facts and Models, Springer-Verlag.
85. Fukunaga K (1990), Introduction to Statistical Pattern Recognition, Academic Press.
86. Sammon, JW Jr. (1969), IEEE Trans. on Computers., C-18(5):401–409.
87. Kohonen T (1997), Self-Organizing Maps, Springer.
88. Torgerson WS (1952), Psychometrika, 17:401–419.
89. Young G, Householder AS (1938), Psychometrika, 3:19–22.
90. Pękalska E, Ridder D, Duin RPW, and Kraaijveld MA (1999), Proc. ASCI'95 (5th Annual Int. Conf. of the Adv. School for Comput. & Imag.), 221–228.
91. Woszczyk W (1982), Proc. of 72nd AES Conv., preprint 1949.
92. Lipshitz S, Vanderkooy J (1981), Proc. of 69th AES Conv., preprint 1801.
93. Thiele N (2001), Proc. of 108th AES Conv., preprint 5106.
94. Bharitkar S, Kyriakakis C (2003), IEEE Wkshp. on Appl. Signal Proc. Audio and Acoust. (WASPAA '97).
95. Radlović B, Kennedy R (2000), IEEE Trans. on Speech and Audio Proc., 8(6):728–737.
96. Bharitkar S, Kyriakakis C (2005), Proc. IEEE Conf. on Multimedia and Expo.
97. Toole FE, Olive SE (1988), J. Audio Eng. Soc., 36(3):122–141.
98. Bharitkar S, Kyriakakis C (2005), Proc. 13th Euro. Sig. Proc. Conf. (EUSIPCO).
99. Talantzis F, Ward DB (2003), J. Acoust. Soc. Amer., 114:833–841.
100. Cook RK, Waterhouse RV, Berendt RD, Edelman S, and Thompson MC (1955), J. Acoust. Soc. Amer., 27(6):1072–1077.
101. Kendall M, Stuart A (1976), The Advanced Theory of Statistics, Griffin.
102. Bharitkar S (2004), Digital Signal Processing for Multi-channel Audio Equalization and Signal Cancellation, Ph.D Thesis, University of Southern California, Los Angeles (CA).
103. Bharitkar S, Kyriakakis C (2000), Proc. IEEE Conf. on Mult. and Expo.
104. Bharitkar S, Kyriakakis C (2000), Proc. IEEE Int. Symp. on Intell. Signal Proc. and Comm. Syst.
105. Nelson PA, Curtis ARD, Elliott SJ, and Bullmore AJ (1987), J. Sound and Vib., 117(1):1–13.

106. Ross CF (1981), J. Sound and Vib., 74(3):411–417.
107. Williams JEF (1984), Proc. Royal Soc. of London, A395:63–88.
108. Elliott SJ, Nelson PA (1993), IEEE Signal Proc. Mag., 12–35.
109. Guicking D (1990), J. Acoust. Soc. Amer., 87:2251–2254.
110. Buckley KM (1987), IEEE Trans. Acoust., Speech, and Sig. Proc., ASSP-35:249–266.
111. American National Standards Methods for the Calculation of the Articulation Index, S3.5-1969 (American National Standards Institute, New York.)
112. Cantoni A, Butler P (1976), IEEE Trans. on Comm., 24(8):804–809.
113. Robinson E (1967), Statistical Communication and Detection, Griffin.
114. Rabiner L, Gold B (1993), Theory and Application of Digital Signal Processing, Prentice-Hall.
115. Makhoul J (1981), IEEE Transactions on Acoust., Speech, and Sig. Proc., ASSP-29:868–872.
116. Yule GU (1927), Philos. Trans. Royal Soc. London, A226:267–298.
117. Söderström T, Stoica P (1983), Instrumental Variable Methods for System Identification, Springer-Verlag.
118. Schroeder MR (1954), Acustica, 4:594–600.
119. Doak PE (1959), Acustica, 9(1):1–9.
120. Childers DG (2000), Speech Processing and Synthesis Toolboxes, John Wiley.
121. Makhoul J (1975), Proc. IEEE, 63(4):561–580.
122. Orfanadis SJ (1985), Optimum Signal Processing, Macmillan.
123. Mourjopoulos J (2000), Room Acoustics Simulator v1.1, Wireless Communications Laboratory, University of Patras, Greece.
124. Gray RM, Buzo A, Gray AH Jr., and Matsuyama Y (1980), IEEE Trans. on Acoust. Speech and Sig. Proc., ASSP-28(4):367–376.
125. Ash RB (1972), Real Analysis and Probability, Academic Press.
126. Hayes M (1996), Statistical Digital Signal Processing and Modeling, John Wiley.
127. Snyder SD (2000), Active Noise Control Primer, Springer.
128. Olson HF (1957), Acoustical Engineering, Van Nostrand.
129. Elliott SJ (2001), Signal Processing for Active Control, Academic Press.
130. Lueg P (1936), Process for silencing sound oscillations, US. Pat. No. 2,043,416.

Index